Mosses, Liverworts, and Hornworts

Pogonatum urnigerum (page 178), North Traveler Mountain, Baxter State Park, ME

Mosses, Liverworts, and Hornworts

A Field Guide to Common Bryophytes of the Northeast

RALPH POPE

Comstock Publishing Associates
a division of
Cornell University Press
ITHACA AND LONDON

First published 2016 by Cornell University Press

First printing, Cornell Paperbacks, 2016

Printed in the United States of America

Library of Congress Cataloging-in-Publication Data

Names: Pope, Ralph, author.
Title: Mosses, liverworts, and hornworts : a field guide to common bryophytes of the Northeast / Ralph Pope.
Description: Ithaca : Comstock Publishing Associates, a division of Cornell University Press, 2016. | Includes bibliographical references and index.
Identifiers: LCCN 2016005759 | ISBN 9781501700781 (pbk. : alk. paper)
Subjects: LCSH: Bryophytes—Northeastern States—Identification.
Classification: LCC QK533.82.N67 P67 2016 | DDC 588.0974–dc23
LC record available at http://lccn.loc.gov/2016005759

Cornell University Press strives to use environmentally responsible suppliers and materials to the fullest extent possible in the publishing of its books. Such materials include vegetable-based, low-VOC inks and acid-free papers that are recycled, totally chlorine-free, or partly composed of nonwood fibers. For further information, visit our website at www.cornellpress.cornell.edu.

Paperback printing   10 9 8 7 6 5 4 3 2 1

Typefaces: Minion Pro (text) and Myriad Pro (display).

Dedicated to Nancy Slack
A constant source of inspiration,
her teaching, writing, and research
provide much of the intellectual
foundation for this book.

Contents

Acknowledgments

Nancy Slack has been my professor of bryology at the Eagle Hill Institute since 2002. She is a skilled bryologist, and her ecological focus puts these organisms in a larger context that provides much of the conceptual basis for this field guide. I am thankful for her comments, her corrections, and most important, her encouragement.

Fellow students at Eagle Hill Institute have also been a great help. I'd like to give special thanks to Ralph Ibe, Bill Miller, and Anne Mills; all offered helpful comments on early drafts.

Robert Ireland's reference books have been my constant companions for so many years that I feel as though I know him. He writes clear and concise keys, and after trying to write a key for this book, I realized he had already written a key far more usable than anything I was likely to produce. The master key and the primary dichotomous keys to the acrocarpous and pleurocarpous mosses and to the liverworts, are highly modified versions of the keys in Robert Ireland and Gilda Bellolio-Trucco's *Illustrated Guide to Some Hornworts, Liverworts and Mosses of Eastern Canada*, a wonderful bryophyte primer published as *Syllogeus 62* by the Canadian Museum of Nature. *Syllogeus 62* provides identification information for most of the bryophytes likely to be encountered in eastern Canada and New England, and because it does so without requiring the use of a compound microscope, it made a fine starting point for this book. I'm indebted to the authors and publisher for making it available. Nomenclature has been updated. It is used here with permission from the Canadian Museum of Nature, Ottawa, Canada.

Bowdoin College has generously allowed the Jossy Cryptos, an informal subset of the Josselyn Botanical Society, to use their laboratory facilities. Our group meets at Bowdoin irregularly in fall and spring courtesy of lab managers Shana Stewart Deeds, Jaret Reblin, and Nancy Olmstead, and professors Nat Wheelwright and Barry Logan.

The Jossy Cryptos provide great feedback and help me keep mosses on my front burner. Alison Dibble and Pat Ledlie are our ringers, with a strong supporting cast, including Eric Doucette, Barbara and Charlie Grunden, Helen

Koch, Joanne Sharpe, Lauren Stockwell, Dorcas Miller, and Jean Wood. Cloe Chunn and Marnie Reeve have been a help, particularly in the world of peat mosses. Additional "occasional" Jossy Cryptos have been a big help as well, and I am grateful also to them.

Finding species to photograph is one of the great challenges, and for help with this effort I'd like to thank Dorothy Allard and Alison Dibble. Susan Munch, author of *Outstanding Mosses & Liverworts of Pennsylvania & Nearby States*, went out of her way to guide me to several of the species in this book.

Dorothy Allard, Keith Bowman, Bill Buck, Eric Doucette, Joe Francis, Janice Glime, Ekaphan Kraichak, Joanne Sharpe, and Stephanie Stuber have all offered valuable editorial comments that greatly improved this book. Feedback from Cornell University Press anonymous reviewers also helped shape this effort.

Jerry Oemig, my go-to person for liverwort expertise, has been very generous with his time and knowledge.

The University of Maine Herbarium has been a great resource thanks to Professor Chris Campbell, Eric Doucette, and Garth Holman. Eric Doucette did a marvelous job producing the distribution maps.

I am indebted to Robert Klips, associate professor of evolution, ecology and organismal biology at Ohio State University at Marion, who contributed several excellent photographs. Hermann Schachner contributed 7 photographs through Wikipedia Commons.

I am also grateful to the Consortium of North American Bryophyte Herbaria and to their contributing institutions.

Cornell University Press editors Kitty Liu and Susan Specter have been tireless in keeping this complex project on track in spite of many obstacles. They have dealt with all issues promptly and with good grace and humor. Special thanks are due to my wife, Jean, always my first reader and kindest critic.

In spite of all this wonderful help, I, of course, take full responsibility for any errors.

I hope readers will feel welcome to send suggestions, corrections, and comments to me at rhpope351@aol.com.

Mosses, Liverworts, and Hornworts

Introduction

The term bryophyte includes mosses, liverworts, and hornworts, all of which are small, generally nonvascular, spore-producing plants. Bryophyte may or may not be a proper taxonomic term, but it is widely used by botanists, and I use it here to include the three groups of plants included in this field guide. Because of their small size and limited morphological and color diversity, mosses, liverworts, and hornworts can be a challenge to identify without a microscope.

The first person to write a reasonably comprehensive macroscopic photo guide to mosses was A. J. Grout in 1900, when he self-published his ground-breaking photo-based *Mosses with a Hand-Lens*. By the 1920s, Grout had added the liverworts and hornworts to complete the bryophyte group. He continued to tinker with the idea of a popular guide to provide "a stepping stone to the larger and more complete works, and to the broader and fuller study of bryology" until his final edition of *Mosses With a Hand-Lens* was published in 1947. Interspersed with his hand lens guides were many other more technical publications such as *Mosses with a Hand-Lens and Microscope*, and his three-volume magnum opus, *Mosses of North America North of Mexico*. More than a century after Grout began his experiment, we're giving the "popular" photo-based field guide to bryophytes another try.

The target audience for this book is hikers, amateur naturalists, nonbryologist botanists, and, possibly, sophisticated gardeners. To get the most from this book, readers should have a 14–20× hand lens, and I hope the book gives them the tools necessary to key to species most of the mosses, hornworts, and liverworts they encounter. If identification to species isn't possible using hand lens characters, we should still consider it a success if this book helps a reader key to a genus or to a species group of some sort that makes sense. If successful, this book will encourage many readers to pursue further study of bryophytes.

Much of the work presented here has been framed by Robert Ireland and Gilda Bellolio-Trucco's *Guide to Some Hornworts, Liverworts and Mosses of Eastern Canada*, published by the National Museums of Canada in 1987. Like Grout's "popular" guides, this publication provides tools for the reader to identify the more common bryophytes without the aid of a compound

1

microscope. This and Ireland's more comprehensive and technical *Moss Flora of the Maritime Provinces* (1982) have been my constant companions as I've studied the mosses along the coast of Maine where I live.

ILLUSTRATIONS AND PHOTOGRAPHS

As did Grout, I have taken advantage of the amazing skills honed by naturalists before the widespread availability of photography. The text and photographs are enhanced by drawings taken from the extraordinarily well-rendered plates of *Bryologia Europaea* (Bry. Eur. in the attributions) and Sullivant's *Icones Muscorum* (Ic in the attributions). Where good existing drawings were not available to Grout, he commissioned illustrations by Mary Thayer (M.V.T.), and I have used many of those.

Except in a few instances noted in the captions, the photographs are by the author. Most were shot with a Nikon digital single-lens reflex camera and Nikon lenses. The close-up shots were taken with a 105 mm Nikon macro lens and at least two strobes. To increase depth of field, many of the photographs are composites of several photographs taken at slightly different focus points, then combined using Adobe Photoshop stacking software. Several excellent photographs by Bob Klips and Hermann Schachner appear in the book and I am grateful for their contributions. Please see the acknowledgments for more information.

COVERAGE

Geographical

The two most important choices in preparing this book were the geographic area to cover and the species to include. I expect that this book will provide reasonable coverage centered on New England, stretching north to Maritime Canada, west to Southern Ontario, and south in the Atlantic States from New England through Maryland. South of Maryland through the mountains of the Carolinas and Georgia, coverage might be reasonable, though it has not been evaluated. West of the Atlantic states, in southern Michigan, Ohio, Indiana, and Illinois, the coverage is likely to be less useful, in large part because of very different geology.

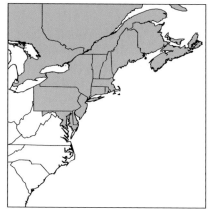

Species

The other coverage decision is the particular species to include. The mosses, liverworts, and hornworts of coastal Maine are essentially the same as those of Maritime Canada, so it made sense for me to begin with the list of species covered by Ireland's nontechnical manual discussed above and to make additions and deletions allowing for my different range. The selection of species in this book is not exhaustive, but it does represent the more conspicuous or unique species that a hiker or amateur naturalist might encounter. I haven't included small inconspicuous species, and I've stayed away from rarities unless I had some compelling reason to include them. *Buxbaumia aphylla* and *Schistostega pennata* are examples of species not frequently encountered but just too much fun to skip. Liverworts, often smaller and less conspicuous than mosses, are treated less intensively—the discussion limited in many cases to representatives of the more common genera. If I've left out your favorite bryophyte, please let me know.

SPECIES NAMES

In most cases, scientific nomenclature conforms to Tropicos (Tropicos.org), a taxonomic database published by the Missouri Botanical Garden. Many common names are used for bryophytes throughout the world, but few are widely accepted, and referring to the species by their scientific binomials is the rule. I have refrained from muddying the nomenclatural waters with common names of my own invention, and have listed only those few names that seem to me to be in general use in our coverage area.

SPECIES RANGES

Ranges for bryophytes are difficult to establish for several reasons: few people are collecting, identifying, and labeling samples for herbaria; there may be a bias toward collecting more unusual or uncommon species, many herbaria have not digitized their collections, and many herbarium specimens do not have GPS data attached. Not surprisingly, the greatest bryophyte diversity tends to appear in places where a bryologist has lived.

Understanding the limitations, helpful distribution information is available from the Consortium of North America Herbaria, which makes available GPS data from more than 50 North American herbaria. This book includes distribution maps based on consortium data. These distribution maps need to be used with an understanding that the underlying data are sparse, and that the actual range of many of the species may be significantly larger than portrayed. That said, the maps do provide a good overview of the general distribution pattern and can be helpful in arriving at a species determination. The maps provided are intentionally small to minimize the implication of precision where there is none. Because liverwort and hornwort data are even more sparse than for the mosses, distribution maps are not provided for those species.

If you wish to access detailed information about collection locations for a particular species, I suggest you begin with the very user-friendly website of the Consortium of North American Bryophyte Herbaria (CNALH), at http://bryophyteportal.org (accessed in September 2015). They continue to add specimens and GPS data at a rapid pace.

GPS data used in creating the distribution maps were published by the following institutions:

Academy of Natural Sciences of Drexel University (PH)
Bell Museum of Natural History, University of Minnesota (MIN)
Boise State University Lichen Herbarium (SRP)
Botanical Research Institute of Texas (BRIT)
Brigham Young University (BRY)
California Academy of Sciences (CAS)
Canadian Museum of Nature Herbarium (CMN-CANM)
Chrysler Herbarium of Rutgers University (CHRB)
Duke University Herbarium (DUKE)
Evergreen Natural History Museum (EVE)
Fairchild Tropical Botanic Garden Herbarium (FTG)
Farlow Herbarium (Harvard University-FH)
Field Museum of Natural History (F)
George Safford Torrey Herbarium (UConn-CONN)
Illinois Natural History Survey (ILLS-ILLS)
Indiana University Herbarium (IND)
Intermountain Herbarium, Utah State University (UTC)
Louisiana State University (LSU)
Michael I. Cousens Herbarium, University of West Florida (UWFP)
Michigan State University (MSC)
Missouri Botanical Garden (MO)
New England Botanical Club (Harvard University-NEBC)
New York Botanical Garden (NY)
North Dakota State University Herbarium (NDA)
Patricia Ledlie Herbarium (LEDLIE)
Pittsburg State University (KSP)
Pringle Herbarium (VT)
Robert W. Freckmann Herbarium at the University of Wisconsin, Stevens Point (UWSP)
Southern Illinois University Herbarium (SIU)
State University of New York—Binghamton (BING)
The University of Arizona Herbarium (ARIZ)

University of Alaska Museum of the North Herbarium (ALA)
University of California—Berkeley (UC)
University of Central Florida Herbarium (FTU)
University of Colorado, Museum of Natural History (COLO)
University of Florida Herbarium (FLAS)
University of Illinois Herbarium (ILL)
University of Maine Herbaria (MAINE)
University of Michigan Herbarium (MICH)
University of Nebraska Kearney (NEBK)
University of Nebraska State Museum, C.E. Bessey Herbarium (NEB)
University of New Hampshire (NHA)
University of North Alabama (UNAF)
University of North Carolina Herbarium (NCU)
University of Richmond Herbarium (URV)
University of South Florida Herbarium (USF)
University of Tennessee Herbarium (TENN)
University of Washington Herbarium (WTU)
University of Wisconsin—Eau Claire (UWEC)
University of Wisconsin—Madison (WIS)
University of Wyoming Rocky Mountain Herbarium (RM)
Valdosta State University (VSC)
Wesley E. Niles Herbarium, University of Nevada, Las Vegas (UNLV)
West Virginia University Herbarium (WVA)
Yale University Herbarium, Peabody Museum of Natural History
(YPM-YU)

BRYOPHYTE BIOLOGY

Bryophytes are not just tiny versions of larger vascular plants. They have very different physical characteristics, and they solve many of life's problems differently from the vascular plants that dominate plant life on Earth.

Bryophytes are small, and small size has allowed them to colonize a great diversity of habitats throughout the globe. Don't allow their small size to fool you into thinking they are evolutionary failures. Their body plan and lifestyle have suited them well, allowing them to survive for more than 400 million years with apparently minor changes. Approximately 20,000 bryophyte species are known worldwide, and in much of the Arctic, they are the dominant life form.

Water and Morphology

While a few families, particularly Polytrichaceae, do have functional vascular systems, Many bryophytes have no vascular system—at least not the complex fluid transport system we've come to understand in larger plants, and they have no woody reinforcement; however, with or without complex plumbing, bryophyte cells all need access to nutrients, gas exchange, and water. With their often minimal piping system, it can be difficult to move water and nutrients throughout the plant, and without woody reinforcement they cannot develop a large supporting structure. Structures need to be small to give cells reasonably direct access to the environment. This is poikilohydry, where the water content inside the plant is directly affected by water available outside the plant. It contrasts with homoiohydry, found in most vascular plants, where a root structure and vascular system maintain hydration inside the plant regardless of water availability outside the plant. Next, let's look at why the minimal bryophyte piping system never evolved to become a tiny version of vascular plants' xylem and phloem.

Simple geometry shows that small shapes (and plants) have a much higher surface-to-volume ratio than larger shapes. The variable in the formula for the surface area of a three-dimensional shape is r^2, while the variable in the formula for the volume of a shape is r^3, so as a plant gets larger, its surface area grows by the square of a dimension while the volume grows by the cube of the same dimension, providing a dramatically lower surface-to-volume ratio for the larger object. This means that larger plants have more internal space, where fluid can be transported, per unit of surface area, where fluid is gained from or lost to the atmosphere.

So, as a result of their much larger volume per unit of surface area, larger plants with woody vascular structures were able to develop sophisticated plumbing leading to cuticle, stomata, deciduous leaf drop, and a host of other homoiohydric adaptations, while the tiny bryophytes evolved their suite of poikilohydric mechanisms to provide water and nutrients to their cells. Each group developed adaptations appropriate to its stature.

Bryophytes are designed to acquire and hold water. The fact that many bryophyte cells need direct access to the environment to be hydrated and to acquire nutrients makes them highly susceptible to desiccation, so these plants have evolved strategies to deal with dry spells. First, much of their structure has evolved to hold water. *Sphagnum* species have specialized cells that are empty and porous at maturity, providing extensive water storage capacity. Many bryophyte species produce delicate structures along their stems and branches (paraphyllia and rhizoids) that provide more surface for capillary water retention. Species in the Polytrichaceae family (also the genus *Aloina* in Pottiaceae, a rare genus in our region) have extra rows of cells on their leaf surfaces (lamellae) that may enhance water storage. Some species' leaves are cupped and held close to the stem, and some are rolled into tubes—both are strategies that draw in water through capillary action.

Some bryophyte groups have evolved leaves that fold up against the stem, or contort and twist during dry spells, providing capillary spaces for water retention, and preventing air from flowing directly over the wet surfaces. Like many alpine plants that are exposed to wind and sun, bryophytes often crowd together in low mounds, minimizing wind exposure, reducing water and heat loss. Remember the old adage that if you happen to lose your compass, your iPhone, your GPS, your ability to see the sun, and your sense of direction, moss growth will show you the north side of a tree? Well, keep the compass handy, but the north side of a tree trunk does indeed get less desiccating sunlight than the rest of the tree trunk, so it just might have more moss growth. Score one for the Boy Scouts. Most bryophytes find shady, damp places to grow, though there are many exceptions.

Many bryophytes are desiccation tolerant. For those species filling less sheltered niches (think tree trunks and rocks), drying out is a part of life. Not surprisingly, many of these species have evolved complex and sophisticated cellular repair mechanisms to enable life to continue even after severe desiccation.

Sexual Reproduction

The green bryophyte plant that we are identifying (the elaborative phase) is the gametophyte, which means it is haploid, that is, with only one set of unpaired chromosomes—n in the illustration on page 8. This contrasts with most life on Earth, which is diploid, having 2 sets of chromosomes—$2n$ in the illustration. The haploid green gametophyte produces male and female sex organs (antheridia and archegonia), which in turn produce the haploid gametes, the sperm and eggs.

Antheridia and archegonia may be produced on the same plant (monoicous) or on separate plants (dioicous). Bryophytes produce biflagellate swimming sperm that need at least a film of water to swim to and fertilize an egg. The sperm have 2 flagellae, take a semicircular shape, and appear to swim in circles. Their movement seems random when viewed under a microscope by the voyeuristic bryologist, but studies show they do follow chemical signals toward the archegonia, a flask-shaped structure housing the ovary and egg. After fertilization, the diploid ($2n$ in the illustration) zygote develops into the sporophyte generation, the diploid generation of the plant. This sporophyte generation (usually capsules), where the spores are produced, is typically not, or only minimally, photosynthetic and receives most of its nutrition from the green, haploid gametophyte.

Spores, broadcast from the capsules by wind, water, insects, other animals, or gravity, are microscopic, quite durable, and capable of traveling great distances in wind currents or on feathers or fur. Some bryophyte species have global distributions.

But it's not all about sex . . .

Asexual Reproduction

Sexual reproduction rather obviously requires both archegonia and antheridia to be in relatively close proximity, and to have water available for the swimming sperm—conditions not always in place, particularly in dioicous species, which have male and female sex organs on different plants. Fortunately, bryophytes have other ways to get the job done.

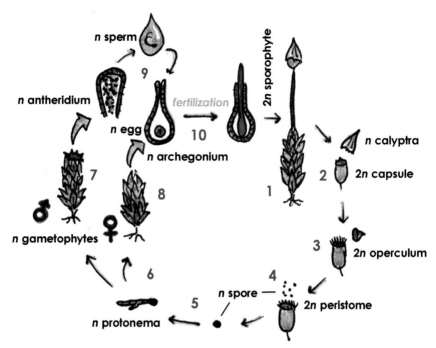

1, A haploid gametophyte produces a diploid (2n) capsule; 2–5, The capsules produce haploid (n) spores; 6–9, A spore grows into a (usually) filamentous protonema that eventually produces the gametophyte plant that includes the archegonium and antheridium (all n); 10, The gametes unite to produce the 2n sporophyte. Thanks to Stephanie Stuber for this illustration, which first appeared in her 2012 book *The Secret Lives of Mosses: A Comprehensive Guide for Gardens.*

Totipotent cells. Most bryophyte cells are totipotent; that is each cell has the capability to grow into a new plant without any sort of fertilization. Herbivory (oddly, not common), trampling, or any sort of shredding can start new plants. This important cell characteristic underlies the ability to produce vegetative propagules.

Moss smoothies. Totipotency allows moss gardeners to establish new colonies by blending moss smoothies. Typical recipes read something like this: In a blender, combine a cup of the target moss species, add a cup of yogurt (makes it sticky? adds nutrients?), a bottle of craft brewed ale (a convenience store brand would never do), and blend coarsely. Buttermilk also appears in many moss smoothie recipes. Spread the resulting green slime on your target substrate, keep the yucky mess damp for a year or so, and new mosses might grace you with their presence. If you're lucky.

Bulbils are small bulblike vegetative propagules produced on the rhizomes and leaf axils of some mosses, particularly *Bryum* species. Each bulbil can detach and begin another plant.

Frangible leaf tips. Several species drop leaf tips as vegetative propagules. See *Dicranum viride* and *Haplohymenium triste* for examples.

Flagellae, or whiplike branches, often break and become vegetative propagules themselves, or they may contain miniature leaves (microphylls) that can break off and grow vegetatively to produce a new plant. *Dicranum flagellare* is an example, though the flagellae are hardly whiplike, and they might better fall under the definition of microphyllous branchlets (below). The liverwort *Bazzania trilobata* reliably produces flagellate branches. See also *Aulacomnium androgynum*, which produces gemmae on flagellate stalks, or *Aulacomnium palustre*, which produces microphylls (tiny leaves) on flagellate stalks.

Gemmae. Many species produce gemmae, which are specialized vegetative propagules. See *Aulacomnium androgynum* for a good illustration of gemmae on a flagellate stalk, and *Tetraphis pellucida* for gemmae in a splash cup (often called a gemma cup), where the gemmae are broadcast by energy from a falling raindrop. Other reliable gemmae producers are *Marchantia polymorpha*, also with gemmae in cups, and *Blasia pusilla*, with powdery gemmae on flask-shaped structures.

Microphyllous branchlets are small branches with tiny leaves (microphylls). They are often produced in leaf axils or among leaves at the tips of stems. As with the other vegetative propagules, several examples can be given, but *Leucodon sciuroides* and *Pseudotaxiphyllum elegans* are good illustrations.

Leaf Morphology

Leaf arrangement. Most mosses have leaves arranged spirally around the main stem, often with 5 positions in 2 turns around the stem (a phyllotaxy of 2/5). Many mosses have a flattened (complanate) aspect, and there are quite a few with leaves spirally inserted around the stem, but pulled into more of a single plane. Leafy liverworts have leaves in 3 rows, usually with 1 row greatly reduced (the underleaves) or missing. Hornworts and thalloid liverworts do not have stems and leaves.

The costa, or midrib, is a thickening of the leaf. It offers reinforcement and sometimes rudimentary vascularity to the leaves, and most moss leaves have some sort of costa. Acrocarpous mosses, the more upright-growing species, have a single costa, usually extending most or all of the leaf length. A costa extending to the tip of the leaf is percurrent, a costa stopping short of the tip is sub-percurrent, and a costa extending beyond the leaf tip is excurrent. Pleurocarpous mosses may or may not have a single costa, often having a double, or forked costa not visible with hand lens magnification. For the purposes of the keys and descriptions in this book, the species with double costae are usually described as ecostate, or lacking a costa.

The leaf lamina is the blade of the leaf and usually 1 cell thick, with any costa and sometimes a border being the only part of the leaf more than 1 cell thick.

Alar cells are the often differentiated cells (frequently enlarged and a different shape than the rest of the leaf cells) in the corners of the bases of leaves where they attach to the stem. These specialized cells control how the leaves react to wet or dry conditions, and technical keys make much use of the shape, color, and number of these cells. See the discussion on differentiated alar cells on page 17.

Growth Forms

Sphagnaceae have a very distinct and recognizable growth form, with a densely branched head (capitulum) and fascicles of usually 4–5 side branches along the stem. Sporophytes are short-lived, simple globular structures that discharge their spores explosively.

Acrocarpous mosses are plants growing in cushions or tufts, with main stems erect or nearly so, simple or occasionally with forked branching, rarely pinnate, and if so, with short tuftlike branches; costae usually prominent and single; sporophytes originating from stem apices, or the apices of major branches. Sporophytes are long-lived; the old ones often present as new sporophytes develop in the following year. Spores are discharged through an opening in the end of the spore capsule, where they are sifted and separated by teeth that frequently surround the capsule mouth.

Pleurocarpous mosses are plants growing in mats, with main stem prostrate or nearly so, sometimes with erect tips or branches, usually much branched, often pinnate, rarely simple, and if so, the stems long and intertwined; costae present or lacking; sporophytes originate from the tip of a very reduced and inconspicuous branch along the main stem somewhere below the apex of a major branch. Sporophytes are as with the acrocarpous mosses.

Hornworts are thalloid, that is, the body of the plant is a flattened plate- or straplike structure with no leaves or stems. Sporophytes are typically tall structures (horns) that discharge their spores by twisting and splitting longitudinally.

Liverworts can be thalloid or with leaves and stems. Some of the thalloid species produce umbrella-like structures or raised plates, or both, supporting the archegonia or antheridia, respectively. If the plant is leafy, the leaves are in two major rows with or without a third row of smaller underleaves. Leaves may be lobed, toothed, or folded into a more complex structure (complicate-bilobed). Sporophytes are short-lived, with a usually black globular capsule that discharges spores by rupturing and splitting apart.

Nutrient Uptake

Cation exchange is a mechanism by which plants acquire nutrients. In cation exchange, a plant releases a proton (H+) into its environment; that proton dislodges and replaces a positively charged nutrient cation from a negatively charged soil or humus site where it has been loosely held, and that cation is now available to bond to a site on a plant cell. Some plant groups are better at this trick than others, and as it turns out, bryophytes are very, very good at cation exchange, allowing them to access nutrients where they are in scarce supply. This ability to access nutrients in acidic, nutrient-poor conditions is what allows *Sphagnum* species to dominate bog systems and arctic tundra where conditions make it difficult for other plants to acquire the needed nutrients. (More on this important characteristic in the Bryophyte Ecology section.)

BRYOPHYTE TAXONOMY

Mosses, hornworts, and liverworts, the three groups making up the bryophytes, evolved from the aquatic ancestors of modern green algae and represent the beginnings of terrestrial plant life, eventually giving rise to our amazingly diverse array of vascular plants.

In the mid-eighteenth century, Carl Linnaeus began the system of scientific nomenclature we use today. This Linnaean system has worked surprisingly well over the past 250 years or so, given that he devised his scheme to illuminate and celebrate God's plan, but we use his classification system today to indicate evolutionary relationships.

Through the years many proposals have been made for bryophyte taxonomy, and spirited discussions continue today. Here's a look at one taxonomic system, somewhat altered to eliminate groups not in our range:

Kingdom Plantae
Subkingdom Viridiplantae (green plants)
Infrakingdom Streptophyta (land plants)
Division Bryophyta (hornworts, liverworts, and mosses)
   Subdivision Anthocerophytina (hornworts)
   Subdivision Bryophytina (moss)
      Class Bryopsida (the "true" mosses)
         Subclass Bryidae (most of our mosses)
         Subclass Buxbaumiidae (*Buxbaumia* and *Diphyscium)*
         Subclass Dicranidae (Dicranaceae species)
         Subclass Funariidae (*Funaria* and *Encalypta*)
      Class Andreaeopsida (*Andreaea* spp.)
      Class Sphagnopsida (the *Sphagnum* mosses)
      Class Tetraphidopsida (*Tetraphis pellucida*)
   Subdivision Marchantiophytina (liverworts)
      Class Jungermanniopsida

Subclass Jungermanniidae (leafy liverworts)
Subclass Metzgeriidae (simple thalloid liverworts)
Class Marchantiopsida
Subclass Blasiidae (*Blasia pusilla*)
Subclass Marchantiidae (complex thalloid liverworts)

This is only one of several treatments being discussed by bryologists today, and I'm sure there is something here for everyone to dislike; however, a couple of points are worth keeping in mind. First, note that our major groups, the hornworts, liverworts, and mosses, are separated at a high taxonomic level, indicating separate lineages from a very early time in the evolution of land plants. Second, the classification Bryophyta, referring to mosses, hornworts, and liverworts, is thought by many taxonomists to be artificial and unsupported by data; however, it has long been used as a descriptor for these three groups, and it is unlikely to drop out of the botanical lexicon anytime soon.

The organization of this book doesn't neatly follow the above taxonomy. I start with the *Sphagnum* mosses, pulling them out of subdivision Bryophytina, honoring their distinct morphology. Then I discuss the rest of subdivision Bryophytina, artificially separating the upright growing plants (acrocarps) from the more branched, mat-forming plants (pleurocarps). Our fourth and final species grouping includes the subdivision Marchantiophytina, the liverworts. Within that section I split the group into the simple thalloid liverworts, subclass Metzgeriidae; the complex thalloid liverworts, class Marchantiopsida; and the leafy liverworts, subclass Jungermanniidae. Subdivision Anthocerophytina, the hornworts, has been included with the liverworts for convenience.

BRYOPHYTE ECOLOGY

Colonizing nutrient-poor situations. As discussed in the Biology section, bryophytes have a very well-developed ability to extract nutrients from nutrient-poor situations, and *Sphagnum* species are an excellent case in point. They are able to colonize and dominate nutrient-poor, often acidic, situations where most vascular plants can't possibly survive. In acquiring nutrients through cation exchange (see Biology section), they throw off many H+ ions, adding to the acidic nature of their surroundings and making life even more difficult for potential competitors. Capturing and holding water as well as they do results in perennially saturated conditions, and the depth of their growth provides a good insulating layer. In *Sphagnum*-dominated peatlands, the low pH combines with cool temperatures and saturated conditions to minimize microbial activity, resulting in incompletely decomposed plant matter or peat. In much of the boreal north, *Sphagnum* moss is the dominant plant genus, and as it holds much of the planet's carbon, it will be an increasingly important research focus in the future.

Erosion and decomposition. Great moisture retention ability allows bryophytes to gain moisture during a rain event, and then to hold that moisture for an extended period of time, effectively tempering erosion and flooding. This same ability to hold moisture provides closely associated fungi with a steady moisture source to do their job as the chief forest decomposers.

Home to many, food for few. As experienced microscopists all know, bryophytes provide a home and a breeding place for a dizzying variety of small to microscopic organisms ranging from arthropods to water bears. Many birds weave bryophytes into their nests for camouflage and insulation, and possibly for antibacterial characteristics. Herbivory, however, is uncommon. The complex carbohydrates in bryophytes are difficult to break down, and bryophytes often produce chemicals that deter herbivory. Capsules, however, with a higher fat content than the vegetative parts of the plant, are apparently more palatable and occasionally grazed.

Widespread colonization. Advanced cell-repair mechanisms combined with moisture retention, small size, and an impressive array of propagules allow colonization of almost unlimited habitats.

N fixation. Many bryophytes have associations with nitrogen-fixing cyanobacteria (see *Blasia pusilla*), providing nutrients for themselves and, through eventual decomposition, for other plants. Which brings us to succession . . .

Succession. Bryophytes are important in colonizing soil or rock and preparing the way for soil formation and the introduction of vascular plants. Illustrated in the photo on page 14 is bare granitic rock undergoing the beginning stages of succession. The black leading edge of the plant blanket is *Gymnocolea inflata*, a leafy liverwort. Next is yellow-green *Sphagnum compactum*, followed by red *Sphagnum capillifolium*, and finally vascular plants. The order of succession occurs in many possible scenarios depending at least in part on the source and location of the barren substrate, but bryophytes are very often part of the mix.

Environmental monitors. Lacking a sophisticated vascular system, most bryophyte cells gain access to water, nutrients, and air through direct contact with the environment. This intimate contact between the bryophyte cells and the environment makes them particularly susceptible to environmental degradation and pollution. In addition, bryophytes are evergreen and lack bark, so they don't have the ability to eliminate potentially poisonous chemicals by dropping leaves and sloughing bark.

Human Uses

Diapers, feminine hygiene, and wound dressing. Antibiotic qualities and the ability to hold up to 20 times their weight in water make *Sphagnum* species excellent for these uses.

Colonizing bare rock on the Welch and Dickey Mountain Loop Trail, Thornton, NH

Fish bait. Frogs, worms, hellgramites, and all manner of fishing bait have long been kept nicely hydrated and alive in fresh *Sphagnum* moss.

Heat. As the vegetation in a *Sphagnum*-dominated bog partially decomposes and compresses under the weight of the next generations of moss, it forms peat, a carbon-rich source of heat used to warm northern homes for all of recorded history.

Insulation. *Sphagnum* again, but see also *Fontinalis antipyretica* (*antipyretica* translates as "against fire"), which was reportedly used in Scandinavian countries to chink chimneys. This was a potentially unpleasant use, given that decaying aquatic invertebrates associated with this species often lend a fishy odor to less than fresh *F. antipyretica* samples.

Medicinal uses. See wound dressing above. Not surprisingly, a group of plants this old and diverse has been used medicinally for many reasons over the centuries; however, many uses have been based on the Doctrine of Signatures, which postulates that God endowed certain plants with the ability to cure diseases and has signaled those properties by making the plant or plant part resemble the organ to be cured. The diversity involved makes it likely that some of the chemicals produced may be found medically helpful in a more scientific way.

Packing material. The densely packed stems and leaves with all the ancillary structures that build capillary capacity have provided reliable packing material for many years.

Pillow and mattress stuffing. The same characteristics that make resilient packing material also make good sleeping material. The genus *Hypnum* (*hypnos* is Greek for sleep) may be so named for providing sleep-inducing pillow stuffing.

Natural pesticides. Many bryophyte species appear to produce natural pesticides, a characteristic important for many of the uses enumerated above, and very helpful in herbaria, where bryophytes don't need the antipest treatments so critical for storage of other plant specimens.

Plant nursery trade. Moss, particularly peat moss, has many uses in the gardening arena. Its water retention capacity makes it wonderful bedding for plants that need steady and controlled moisture, and it is used as an amendment to condition and aerate soil.

Scotch whisky. Most important of all, Scotch whisky (from the Gaelic word *uisce*, meaning water of life) with its smoky, peaty flavor, is made from bog water and barley that has been moistened, allowed to sprout, and then dried over an open peat fire. We are fortunate indeed that much of Scotland is covered by peat.

Bryophyte ecology is a large and growing area of study, and I encourage you to explore the subject beyond the few high points touched on here. *Bryophyte Ecology and Climate Change*, by Zoltan Tuba et al., and Janice Glime's online book *Bryophyte Ecology* are excellent places to launch a more in-depth look at this subject. Both resources are discussed in the annotated references.

COLLECTING AND STUDYING BRYOPHYTES

First, be sure you have a bryophyte. Lichens can, and often do, share habitat with bryophytes. Lichens are dual organisms, combining a fungus with an algal or cyanobacterial component. Like bryophytes, they are small, low growing, and spore producing, and they can be found in a wide range of habitats. Unlike most bryophytes, lichens never have stems and leaves, and when dry, lichens are yellow-green or gray-green to darker, unlike the more grass-green bryophytes. Also, bryophytes often produce capsules that may be held above the plant on a stalk, or may be embedded in leaves. A bit of page flipping through this book will give you the feel of what these capsules look like. The thalloid liverworts and the hornworts, lacking stems and leaves, are the most likely to be confused with lichen. The presence of specialized reproductive structures (see Liverworts and Hornworts), and color will help separate the groups. Some species of bryophytes may grow submerged, while lichens almost never do. Small vascular plants can be particularly confusing, but bryophyte leaves are usually only 1 cell thick whereas the leaves of vascular plants are much thicker.

Collecting. When you find a specimen you would like to collect and identify, take a minute to look around to ensure you're getting a good, representative example, with sporophytes if available. Don't skimp. Collect a sample the size of your palm, assuming you aren't denuding a small colony. Most bryologists collect in small paper bags, recording on the bag the date, location, GPS coordinates, substrate, overall habitat, elevation if significantly above sea level, and a possible name, if you've got an idea of what it might be.

Many of the species in this book can be identified in the field, at least to genus, but it's often helpful to bring your samples home and do your identification at a kitchen table with good lighting and no mosquitoes.

Hand lens/binoculars. A hand lens of 10–20× works well in the field. Lighted lenses can be a great help in dark forests. I personally use Bausch and Laumb 10× and 14× lenses and an Iwamoto 20×, but many options are available, and you will find a great array of hand lenses offered online. Close focusing binoculars can give you a quick view of the plant communities at your feet, and they can save a lot of bending over. Again, many options are available out there, but the Pentax Papilio model focuses to less than 20 inches, it costs around $100, and seems to work well.

A dissecting microscope with at least 20× magnification will be a great help in your work. These are available for pretty much whatever you want to spend, depending on magnification options. Dissecting scopes with variable magnification are easiest to use, but not necessary. Variable magnification from 10× to 60× is nice, but expensive.

A compound microscope, typically with magnification from 40× to 1,000×, isn't necessary to use the keys in this book, but will be a valuable and necessary tool if you find bryophytes adequately compelling to work with a more technical manual. Bryophyte slides can be very easy to make—often requiring just a leaf in a drop of water with a cover slip on top—and they reveal a beautiful and highly varied cellular structure.

Examining the specimen takes some patience. Those of us used to identifying vascular plants may have expectations of quick and accurate determination without a lot of study, but it doesn't always work that way with bryophytes. Of course we get to know some of our local favorites at a glance, but most identifications require some time for careful observation and keying.

Differentiated Alar Cells

The bottom corners of moss leaves are referred to as the alar regions, and in many moss species the cells in the alar region are different in size, shape, or color from the rest of the leaf cells. These differentiated alar cells, located where the leaves attach to the stem, alter the orientation of the leaf as the cells gain or lose moisture. This arrangement can assist the plant in dealing with environmental moisture variation by holding the leaves tightly against the stem in dry spells, or opening the leaves to facilitate photosynthesis when conditions are appropriate. The presence or absence of differentiated alar regions is important in most dichotomous keys, including the ones in this book.

As you work your way through a key you have two choices when faced with a question regarding presence or absence of differentiated alar cells. One choice is to follow both sides of the key and pick the answer that seems most appropriate; however, opportunities for wrong choices in a dichotomous key can be abundant, and examining the alar cells in most species just isn't that difficult.

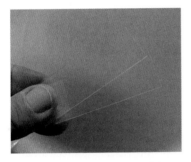

1. Begin with two microscope slides.

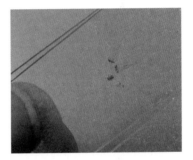

2. Using fine tweezers, pluck two or three leaves from your moss stem and place them on a slide in a drop or two of water. Put the second slide on top of the leaves to hold them flat.

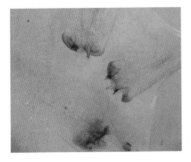

3. Now you're ready to hold the slide up to the light to examine the alar cells with a 10–20× hand lens. In most instances, no compound microscope is required to see whether the alar cells are differentiated.

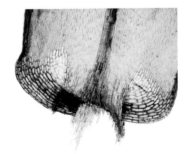

However, if you do have access to a compound microscope, you can uncover a beautiful new world! These alar cells are differentiated in shape, size, and color.

The leaf shown here is from a *Dicranum* sp.

Note the substrate, habitat, and color of your sample. Tease apart the clump to look at individual plants so you can see general size, branching patterns, and sporophytes, and find detail structures such as paraphyllia, rhizoids, gemmae, flagellate branches, and the like.

Examine leaves for orientation with relation to the stem (appressed, divergent, squarrose, falcate-secund, etc.), presence or absence of a costa, general shape (ovate, lingulate, lanceolate, acuminate, etc.), note whether leaves are cupped, flat, keeled (folded), with margins incurved or recurved, whether leaves have a noticeable margin, whether leaf arrangement is spiral around the stem, or whether leaves have grown to give branches a flattened appearance (complanate) or are actually inserted in two or three rows (uncommon in mosses, quite common in liverworts). Also note teeth, or lobes (lobes are liverworts only), and whether leaves have more than a single face, that is, whether they have a folded over/under flap (*Fissidens* spp. in mosses, otherwise, liverworts), and whether that secondary flap might be shaped into a structure other than a flat leaf surface such as a cup or globe shape. Also note the different shape taken on by leaves as they dry.

The dichotomous moss keys ask in three places whether differentiated alar cells are present. See the discussion of differentiated alar cells on page 17.

Sporophytes, when present can be very helpful. Note whether the seta holds the capsule well above surrounding leaves or if the capsule is immersed in leaves. Note how many setae originate from the same place (usually just 1 seta per archegonia), whether the seta originates from the tip of the main stem or major branch (acrocarpous) or originates from a small and often inconspicuous branch at a point well below the top of the plant (pleurocarpous). What color is the seta and is it smooth or rough, and is it twisted? Note the capsule size, general conformation, and the shape of the calyptra, the covering of the developing capsule. If available, note the shape of the operculum (structure covering the capsule mouth), and the number of teeth.

When you get your specimen home, make a list of the answers to the questions as you work through the key, and include those notes with your specimen.

Making packets. Bryological specimens are stored in flat paper packets approximately $10 \times 15\,\text{cm}$ ($4 \times 6\,\text{in}$.). Packet dimensions and folding instructions are shown on the opposite page. Use acid-free paper for best longevity. You can either set up a simple word processing program to print the data on the packet sheet, or use blank paper and print labels separately. A sample label shows the information that needs to be collected.

Before putting the samples in the packets, it's important to flatten them lightly so they aren't broken in the packets. If your samples have dried out, remoisten them and flatten between paper towels.

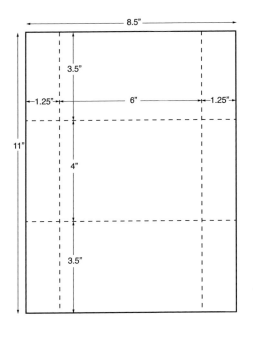

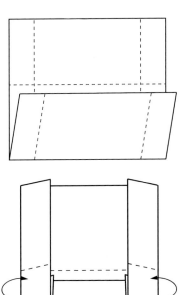

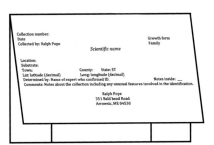

Folding a packet

Clockwise from top left:
1. Layout for a 4 × 6-inch packet
2. Bottom folds up
3. Sides fold in
4. Top folds down
5. Sample label

Collection number:
Date
Collected by: Ralph Pope

Growth form
Family

Scientific name

Location:
Substrate:
Town: County: State: ST
Lat: latitude (decimal)   Long: longitude (decimal)
Determined by: Name of expert who confirmed ID.   Notes inside: __
Comments: Notes about the collection including any unusual features involved in the identification.

Ralph Pope
351 Bald head Road
Arrowsic, ME 04530

Storage. Shoe boxes (women's sizes are best) are often just the right size for storing packets, and they have the advantage of being free.

ORGANIZATION OF THE BOOK

Keys

We have all heard some variation of the sentiment that the only people who can use keys are those who don't need them. Keys certainly can be challenging, but some mechanism to narrow a large field of possibilities down to a manageable group of species is at the heart of any identification book.

Our identification process starts with the master key at the end of this section, the purpose of which is to get you to the general divisions of the book. As with any of the keys, a little page flipping to check results should let you know if you've survived this first test. Good luck.

Once you arrive at the correct division of the book, another key will help you find the correct species page. Various keys differ in style, so you're sure to find something you like (or don't) sooner or later. The *Sphagnum* section begins with a dichotomous key that sends you to a subgenus (section) key, where you encounter other opportunities to make a mistake as you work your way through the section information. Near the front of the *Sphagnum* tab you will find a pair of "quick look" keys that give you a second chance if you run into trouble with the primary dichotomous key. The rest of the mosses are split into two large groups (acrocarps and pleurocarps), and the master key should help you get to the correct tab. Each group has a key at the beginning of its section, with quick keys, subkeys, or charts at the family or genus level. The liverwort and hornwort tab has its own diversity of keys. The hornworts are simply described, without a key (only two hornworts are included in the book), and the liverworts are keyed with a dichotomous key, charts, and a random-access key.

It's not as hard as it sounds. Be sure to collect a good-sized sample, and take your time looking at all the details with as much magnification as you can get. I usually carry 14× and 20× hand lenses in the field. A good dissecting microscope on your desk is invaluable, and if you plan to pursue bryology seriously, you'll need a compound microscope sooner or later.

Much is made in keys, both the macroscopic ones used here and the more technical keys requiring compound microscope work, of the characteristics of the capsules, the sporophyte generation. Unfortunately, capsules, the sporophyte generation, aren't always present; indeed, in some species they are produced rarely, or not at all. We've made efforts here to provide identification clues not requiring capsules, but it's a challenge.

Species Page Arrangement

Sphagnaceae, the peat mosses, are presented alphabetically by subgenus (known as sections in some other references). Acrocarpous and pleurocarpous mosses are presented alphabetically by family. Liverworts and hornworts are presented alphabetically within four groups: the hornworts, simple thalloid liverworts, complex thalloid liverworts, and leafy liverworts.

Species Page Information

Each species page gives the following information, from top to bottom:

Scientific name. Nomenclature is from Tropicos (see websites in annotated references) and other sources.

Photograph(s) showing pertinent identification information with a size bar or other size reference on at least one photograph. The exceptions are the *Sphagnum* pages, which do not have size bars but a life-size photograph of each species.

Species description. Bryophytes have notoriously variable morphology, but I provide color, size, and description here for what appears to be the most common expression.

Similar species. Keying bryophytes is challenging, and this section is an attempt to help separate a species or species group from similar-looking bryophytes.

Range and habitat. See the discussion on species ranges in this chapter. As is true with the range information, the habitat information provided is not meant to exclude other possibilities; it indicates where the species is most commonly found.

Name. The meaning of a scientific name can occasionally provide insight into characteristics of that species and often provides a useful mnemonic. Sometimes the authors of species names make their intentions clear and obvious, but sometimes the intentions are cryptic or obscure. When a name is originally published, the author of the name provides a species description, traditionally in Latin but now may be in English, and that description (called a protologue) often gives clues to the reason for the name; however, many of these protologues either don't address the name choice or are not readily available, so a certain amount of poetic license has been necessary in interpreting names and their meaning. If there is a common name that appears to be in wide use, it is included here.

Family. See the introduction. In the *Sphagnum* section, the name at lower left on each species page is the subgenus (section), since all the *Sphagnum* spp. in this book are in family Sphagnaceae.

Ca. If a species favors a calcareous habitat, that is, a habitat with higher than normal pH, "Ca" is shown at the bottom of the page. For liverworts and hornworts, the discussion of pH preference is in the Range and Habitat section.

Distribution maps. Provided for all *Sphagnum*, acrocarpous, and pleurocarpous species. See the introduction for a discussion of these maps.

Abbreviations

Bry. Eur.	*Bryologia Europaea*. 1836–1851. P. Bruch, W. P. Schimper (ed.), and T. Gümbel. Sumtibus Librariae E. Schweizerbart. Stuttgart. Used here as a source of illustrations.
Ca	At the bottom of a page is my shorthand indicating that a species has a preference for calcium enrichment, reflected in higher pH. Within the range of this book, higher pH is commonly associated with rock containing calcium carbonate, usually limestone or marble; however, other minerals such as basalt can cause the higher pH environment indicated by the Ca symbol.
G	Greek
Ic	*Icones Muscorum*. 1864. William Starling Sullivan. Cambridge, MA: Sever and Francis. Used here as a source of illustrations.
L	Latin
M. V. T.	Mary V. Thayer, illustrator of many reference books by A. J. Grout.
R	A drawing by the author.
sp.	Following a genus name, a single unknown or unidentified species in that genus.
spp.	Following a genus name, more than one species in the genus.
subg.	Subgenus, used in the Sphagnaceae chapter to delineate subgroups within that family that are often referred to as sections.
subsp.	Following a species name, a subspecies.

MASTER KEY

1. Plants thalloid..2
1. Plants leafy...3

2. Plants circular in outline or nearly so, thin, dark-green, without upper surface markings or costa (midrib); capsules long-cylindrical, lacking a seta (stalk), splitting longitudinally...**Hornworts**
2. Plants not usually circular, often thick, light to dark green, sometimes with upper surface markings and midrib; capsules not long-cylindrical, and often with some sort of stalk... **Thalloid Liverworts**

3. Leafy plants with 2–3 ranks of leaves (one row on each side of stem, and a third, if present, small, and on the underside of the stem midway between the lateral leaves), the leaves usually round, lobed, or deeply incised; costa (midrib) lacking; capsules opening simply by splitting into 4 valves.................
....**Leafy Liverworts***
3. Leafy plants with leaves arranged spirally around the stem, the leaves rarely round, or, if so, never lobed or deeply incised; costa (midrib) often present; capsules complex, usually with a rim of teeth around the opening 4

4. Plants with 3 to several branches in fascicles (bunches), the branches crowded near the stem tip to form a tuft; in bogs, swamps, lakes, wet depressions in woods, or other wet habitats................................... **Sphagnaceae**
4. Plants without branches in fascicles, often in dry habitats 5

5. Plants in cushions or tufts, main stems erect, or nearly so, simple or occasionally with forked branching, rarely pinnate, and if so, with short tuft-like branches; costae prominent and single; with sporophytes rising from stem apices, or the apices of major branches ....**Acrocarpous Mosses**
5. Plants in mats, main stem prostrate or nearly so, sometimes with erect tips or branches, usually much branched, often pinnate, rarely simple, and if so, the stems long and intertwined; costae present or lacking; with sporophytes rising from the tip of a very reduced and inconspicuous branch along the main stem somewhere below apex of a major branch...
....**Pleurocarpous Mosses**

Fissidens and *Distichium* are moss genera with two-ranked leaves that might key as a Leafy Liverwort. *Distichium* is not common, but *Fissidens* is frequently encountered. See page 137.

Sphagnum torreyanum with capsules, Addison, ME. See page 53. The color of *Sphagnum* capsules varies from the red shown here to dark brown and black.

Sphagnaceae
The Peat Mosses

Many plants can form peat, but in northeastern North America, *Sphagnum* species are the primary peat formers, and these species are often referred to as peat mosses.

Sphagnum recurvum with branch tips darkened by the presence of antheridia in the leaf axils.
See page 55.

SUBGENERA AND SPECIES INCLUDED

The order followed here, alphabetically by subgenus, reflects the order of the subgenus and species accounts. Species in [brackets] are mentioned in keys, but not illustrated.

subgenus *Acutifolia* (p. 37)
 [*S. bartlettianum*]
 S. capillifolium
 S. fimbriatum
 [*S. flavicomans*]
 S. fuscum
 S. girgensohnii
 [*S. quinquefarium*]
 S. rubellum
 S. russowii
 S. subtile
 S. warnstorfii
subgenus *Cuspidata* (p. 48)
 S. cuspidatum
 S. majus
 S. pulchrum
 S. torreyanum
 recurvum group
  S. fallax
  S. angustifolium
  S. flexuosum
  S. recurvum

subgenus *Polyclada* (p. 56)
 S. wulfianum
subgenus *Rigida* (p. 57)
 S. compactum
 [*S. strictum*]
subgenus *Sphagnum* (p. 59)
 S. centrale
 S. imbricatum s.l.
 S. magellanicum
 S. palustre
 S. papillosum
subgenus *Squarrosa* (p. 65)
 S. squarrosum
 [*S. teres*]
subgenus *Subsecunda* (p. 67)
 S. pylaesii
 S. subsecundum s.l.
 [*S. contortum*]
 [*S. lescurii*]
 S. missouricum

IDENTIFYING SPHAGNACEAE

This single genus family, with approximately 40 species in our area and more than 100 in North America, is not large but can be quite unwieldy. Traditionally, the group is separated into subgenera (often referred to as sections) that share some morphological characteristics. This section keys to 35 species and illustrates the 27 *Sphagnum* mosses most likely to be encountered. They are organized in traditional subgenera since this system makes the most sense in a key, and it will help those who wish to pursue a more in-depth study of Sphagnaceae.

As with many mosses, a microscopic examination is required for positive identification of many of these species. However, experienced *Sphagnum* specialists are quite able to make reasonably reliable identifications in the field, so we'll try to ferret out some of the identification clues that they use to do that.

First, let's look at a representative peat moss with a focus on those characters that might help identify the species in the field.

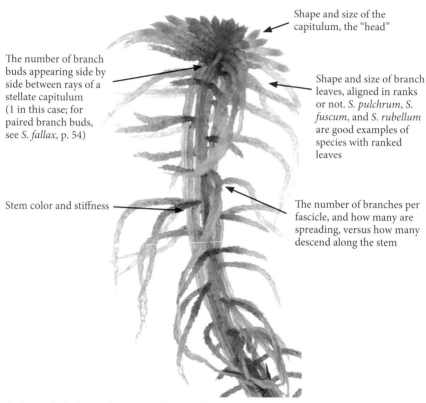

Shape and size of the capitulum, the "head"

The number of branch buds appearing side by side between rays of a stellate capitulum (1 in this case; for paired branch buds, see *S. fallax*, p. 54)

Shape and size of branch leaves, aligned in ranks or not. *S. pulchrum*, *S. fuscum*, and *S. rubellum* are good examples of species with ranked leaves

Stem color and stiffness

The number of branches per fascicle, and how many are spreading, versus how many descend along the stem

S. girgensohnii, shown above, provides a good example of a stellate capitulum with 1 branch bud between rays. It has pointed leaves, not in ranks, usually 2 spreading branches and 2–3 descending branches per fascicle (often written 2+2–3), and a stiff green stem that snaps like celery.

One of the trickier and more important skills is finding and examining the frequently transparent stem leaves, since they are usually quite different from the branch leaves, and their shape and orientation is important to species identification. It helps to let them dry a bit to lose some transparency, or even to blacken them with a marker to make them more visible.

Removing the capitulum will often reveal some stem leaves around the top of the stem, as shown (enhanced).

Plucking branch fascicles off also reveals some stem leaves. Note that these are mildly divergent and growing downward.

As with many of our more aquatic mosses, Sphagnaceae can show striking diversity in color and form within a species depending on environmental factors. For example, species described here as having a color, for example, red for *S. magellanicum*, or brown for *S. palustre*, may be green, particularly when in shade. And species described as having a round capitulum may, from time to time, have a stellate, or at least partly stellate, or flat-topped capitulum. This morphological diversity limits the amount of identification one can reliably undertake in the field. That said, a number of species can be identified in the field under most conditions (*S. magellanicum* is a good example), and many can be identified at a confidence level of 80 percent most of the time. A final determination will often require compound microscopic analysis beyond the scope of this book. The microscopic characters are fascinating, and one of the more salient characters is illustrated below. The green, long, narrow oval cell that houses the photosynthetic apparatus is shown below adjacent to a larger hyaline cell that holds water. The shape and location of these green photosynthetic cells in relation to the hyaline cells is important in the species diagnosis.

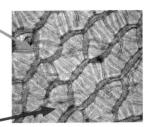

Leaf lamina, flat, *S. palustre*. Chlorophyllous "green" cells (green arrow) surrounded by larger hyaline cells (red arrows).

Leaf cross section, *S. magellanicum*. Chlorophyllous "green" cell, oval-shaped and exposed equally on top and bottom surface of the leaf.

These illustrations are provided to give a general understanding of the nature of microscopic examination of these species, and you are encouraged to see some of the more technical guides discussed in the annotated reference list.

DICHOTOMOUS KEY TO SUBGENERA OF *SPHAGNUM*

It is important to understand the limitations of macroscopic identification of *Sphagnum* species. Some *Sphagnum* species can be identified in the field some of the time. Color is important for field identification, but all species may occasionally be green. Also, as with many plants, these species don't always follow the habitat rules we assign to them. You may need to use all three keys. The first is a traditional dichotomous key to subgenera. Most subgenera have a page that can help you select a likely species name, though in many cases getting to the appropriate subgenus might be considered a good accomplishment.

The second and third keys are "quick looks" that may be helpful with some species some of the time. Use all three keys and page-flip when choices are difficult.

1. Large plants with hooded branch leaves, not squarrose.................................
   .... subg. *Sphagnum* (p. 59)
   > (Subg. *Rigida* plants, which may occasionally key here, have stem leaves ≈0.5 mm
   > long, versus 1.5–2.0 mm long in subg. *Sphagnum*)
1. Not as above 2

2. Densely packed plants with individual plants not easily discernible; often in a
   succession community on wet rock or sand......................... subg. *Rigida* (p. 57)
   > (*S. pylaesii*, in subg. *Subsecunda*, may also key here)
2. Possibly in dense mats, but individual plants clearly visible, or not in a
   succession situation 3

3. Large green plants with leaves squarrose wet or dry...
   .... subg. *Squarrosa* (p. 65)
3. Not as above 4

4. Large plants with more than 6 branches per fascicle; in rich forest sites ....
   .... subg. *Polyclada* (p. 56)
4. Plants with fewer than 6 branches per fascicle 5

5. Plants with few branches, one branch per fascicle; stem leaves much larger
   than branch leaves; minimal capitulum ....
   .... *S. pylaesii* in subg. *Subsecunda* (p. 67)
5. Plants with many branches, more than one branch per fascicle; branch leaves
   as large as or much larger than stem leaves; an obvious capitulum 6

6. Plants with a clearly stellate capitulum..7
6. Plants with a well-developed, but not clearly stellate, capitulum......................8

7. Single emergent branch buds between capitulum rays..
   ....subg. *Acutifolia* (p. 37)
7. Paired emergent branch buds between capitulum rays ....
   .... *S. recurvum* group in subg. *Cuspidata* (p. 48)

8. Plants small to medium-sized; branch leaves <2 mm long, inrolled and with a
   straight, sharp-pointed look; hanging branches longer and thinner than
   spreading branches; stem leaves mostly growing upward and closely
   appressed to stem; stem and branch leaves of similar size; plants frequently
   red; in various habitats, but when in a bog, usually above water level ....
   ....subg. *Acutifolia* (p. 37)
8. Plants small to large; branch leaves various lengths, frequently curved;
   hanging branches longer and thinner than spreading branches or similar;
   stem leaves divergent to divergent-pendent; green or with yellow-brown tint,
   not red; in seasonally wet to inundated situations 9

9. Leaves 1–2 mm long, often curved and with branches upturned in capitulum, not twisted or with wavy margins when dry; hanging and spreading branches similar in length; capitulum green to yellow-brown; lower parts of plants with a bluish tint and shiny when dry; frequently in shaded, seasonally inundated situations .... subg. *Subsecunda* (p. 67)

9. Leaves more than 2 mm long, often curved, mildly twisted and with wavy margins when dry; hanging branches longer and thinner than spreading branches; capitulum green to yellow-green to brown; plants neither shiny nor with bluish tint when dry; frequently in exposed, inundated situations .... subg. *Cuspidata* (p. 48)

Each subgenus page provides additional keys or descriptions. Reviewing the appropriate subgenus pages can give a good sense of the look and feel of the group. For size reference a life-size photo is provided for each illustrated species.

QUICK LOOK BY HABITAT

Not all species are included.

Many species can be found in several habitats. In the following key, species are listed by the habitats that appear to be their favorites. None of the placements is exclusive.

Low-nutrient ombrotrophic bogs or poor fens, at or near water level

subg. *Cuspidata*

S. *cuspidatum*: yellowish green, frequently inundated, limp

S. *majus*: oily look, bog margins

S. *pulchrum*: golden brown, floating lawns

S. *recurvum* group: green, other than S. *fallax*, these are wet specialists

subg. *Sphagnum*

S. *magellanicum*: big, red, low hummocks and bog margins

Low-nutrient ombrotrophic bogs or poor fens, hummock formers

subg. *Sphagnum*

S. *imbricatum*: big, orange-brown, tightly packed, hummocks

S. *magellanicum*: big, red, low hummocks and bog margins

subg. *Acutifolia*

S. *rubellum*: red, often stellate, 5-ranked, hummock former

S. *capillifolium*: red, "pom-pom" capitulum, not 5-ranked, hummock former

S. *fuscum*: brown, often stellate, 5-ranked, hummock top

Medium to rich fens

subg. *Acutifolia*
> *S. warnstorfii*: red

subg. *Sphagnum*
> *S. papillosum*: brown, stubby, cigar-shaped branches

Inundated in shaded site

subg. *Cuspidata*
> *S. torreyanum*: green, may favor slowly moving water

Low-nutrient damp forest sites or shaded margins of shrub swamps

subg. *Acutifolia*
> *S. fimbriatum*: green, very fertile, diagnostic stem leaves
> *S. girgensohnii*: green, single hanging branch, snaps like celery
> *S. russowii*: speckled red

subg. *Cuspidata*
> *S. fallax* in *S. recurvum* group: green, leaves very recurved when dry

subg. *Sphagnum*
> *S. magellanicum*: usually some red
> *S. palustre*: green (in spring) to brown

subg. *Squarrosa*
> *S. squarrosum*: green, favors conifer sites

subg. *Subsecunda*
> *S. subsecundum*: green, curved leaves

Mesic forest sites

subg. *Acutifolia*
> *S. subtile*: red, also in damp sites, not in bogs

subg. *Polyclada*
> *S. wulfianum*: may prefer rich sites

subg. *Squarrosa*
> *S. squarrosum*: green, favors conifer sites

Rich forest sites

subg. *Polyclada*
> *S. wulfianum*

subg. *Sphagnum*
> *S. centrale*: white-green
> *S. papillosum*: brown, stubby, cigar-shaped branches

subg. *Squarrosa*
> *S. squarrosum*: green, favors conifer sites

Seepy rock, succession sites

subg. *Subsecunda*
>*S. pylaesii*: dark reddish brown, wet rock pockets

subg. *Rigida*
>*S. compactum*: light green to yellow-brown, densely packed

QUICK LOOK BY SPECIAL CHARACTERISTICS

Not all species are included. Species in [brackets] are not illustrated.

Hooded branch leaves

subg. *Sphagnum*

Red with hooded branch leaves

Sphagnum magellanicum

Red without hooded branch leaves

subg. *Acutifolia*

Prominent apical bud

Other species may have noticeable, but small apical buds.

subg. *Squarrosa*
>*S. squarrosum*
>[*S. teres*]

subg. *Subsecunda*
>*S. pylaesii*
>[*S. platyphyllum*]

subg. *Acutifolia*
>*S. girgensohnii*
>*S. fimbriatum*

subg. *Cuspidata*
>*S. riparium*

Strongly ranked leaves

subg. *Acutifolia*
>*S. fuscum*
>*S. rubellum*
>[*S. bartlettianum*]
>[*S. quinquefarium*]
>*S. warnstorfii*

subg. *Cuspidata*
>*S. pulchrum*

Single branch buds between rays of stellate capitulum

subg. *Acutifolia*

Paired branch buds between rays of stellate capitulum

subg. *Cuspidata*
  recurvum group

Leaves squarrose when moist

subg. *Squarrosa*
  S. squarrosum
  [*S. teres*]

Leaves squarrose when dry

subg. *Cuspidata*
  recurvum group

Sphagnaceae Subgenera and Species Accounts

Presented alphabetically by subgenus

Subgenus *Acutifolia*

Size Small to medium-sized, large for *S. girgensohnii*, slender.

Ecology Forming compact mats and hummocks in relatively dry acidic habitats (*S. warnstorfii*, a medium to rich fen species, is the exception). If in a bog, these species form hummocks, not lawns.

Pigment Most are red, but brown and green members occur as well. As with all Sphagnaceae, these plants are more likely to show their characteristic pigment when mature, and in sun.

Capitula Well developed, whether stellate and 5-parted, or hemispherical in a "pom pom" shape. If stellate, there will be 1 branch bud between rays.

Branch Leaves Small, generally 1.5 mm long or less, ovate to ovate-lanceolate, entire, with inrolled margins near the apex, giving the leaves a straight, pointed look. Erect when moist to occasionally erect-spreading when dry.

Stem Leaves Large, usually similar to the branch leaves in length, growing upward, appressed to stem.

Name *Acutifolia* is for the sharp-pointed look of the branch leaves.

Microscope Tip In this subgenus, a cross section of a branch leaf will show green cells more exposed on the inside (concave side) of the leaf than the outside.

Key to Illustrated Species of Subgenus *Acutifolia*

1. Plants green throughout, always stellate 2
1. Plants not green, may be stellate or not...3

2. Plants with prominent apical bud; stem leaf as wide as long with lacerate margin extending around the apex and down the sides...............*S. fimbriatum*
2. Plants with prominent apical bud; stem leaf almost rectangular, much longer than wide, with truncate, lacerate tip ....*S. girgensohnii*

3. Plants with brown pigment, if in a bog, usually at top of a bog hummock....... ....*S. fuscum*
3. Plants red, or at least with some red pigment...4

4. Plants in a medium to rich fen.. *S. warnstorfii*
4. Plants of ombrotrophic bogs or on forest floor 5

5. Plants of the forest or forest edge, not in bogs...6
5. Plants not in forest, frequently in ombrotrophic bogs..7

6. Speckled red with lingulate, round-tipped stem leaves 1.2–1.5 mm long........... ....*S. russowii*
6. Solid to partly red, but not speckled, with triangular, pointed stem leaves <1.2 mm long.. *S. subtile*

7. Branch leaves clearly aligned in 5 rows, usually stellate capitula.......................
....*S. rubellum*
7. Branch leaves not aligned in 5 rows, "pom-pom" capitula ....*S. capillifolium*

Possibilities Not Illustrated

S. bartlettianum: Green to red; 5-part capitulum; stem leaves ≈2.0 mm long; branch leaves sharp-pointed and 5-ranked; a bog species. Southern and coastal distribution.

S. quinquefarium: Green to partial pink; stellate to convex "pom pom" capitulum, 5-ranked branch leaves; stem leaves 1 mm long, pointed; 3 spreading branches per fascicle giving a dense look to the individual plants. Conifer forest habitat.

S. flavicomans: Yellow-brown; robust hemispherical capitulum; branch leaves not much ranked; stem leaf resembles *S. capillifolium*; coastal bogs and poor fens from Delaware to the Canadian Maritimes. See comparison, page 42.

An alpine population in full sun, with *Polytrichum strictum*, Mount Mansfield, VT; right, actual size

Description Green to very red with small tight "pom pom" capitulum; branch leaves ≈1.5 mm long, upright to spreading, with recurved margins and a sharp-pointed look, not 5-ranked; upper part of stem may be green, red in lower parts.

Fascicles 3–4 branches, with 2 spreading.

Stem Leaf ≈1.8 mm long with involute margins near the tip, and a somewhat truncate tip (see below left).

Similar Species *S. subtile* is very similar, with red pigment and the "pom pom" capitulum, but is not found in bogs. Habitat and stem leaf length are commonly used to separate the species. See *S. subtile* Similar Species. *S. rubellum*, also red and a bog species, is more likely to have a stellate capitulum, has branch leaves in rows (5-ranked), and occupies wetter sites.

Range and Habitat Throughout our range; typically forming hummocks in and on edges of ombrotrophic bogs, also alpine (top photo). Usually on a firm base, i.e., not on floating bog mats.

Name *capillus* (L) = hair, and *folium* (L) = leaf. The name may refer to the very pointed-looking leaves common throughout this subgenus.

capillifolium

subg. *Acutifolia*

Above, in a forested wet depression; right, actual size; below right, showing the prominent apical bud

Description Slender green plants with a small, stellate capitulum; single branch buds between rays and a prominent whitish apical bud (below right); branches long and graceful; branch leaves acute with recurved margins, ≈1–2 mm long, somewhat reflexed when dry; stem light green. A reliable producer of capsules in early summer.

Fascicles 4 branches, with 2 spreading.

Stem Leaf Broadly fringed across the top and down the sides; ≈1 mm long, and as wide as long. See right.

Similar Species *S. girgensohnii* is the other green member of the subgenus. See Similar Species on that page.

Range and Habitat Greenland to Alaska, south in the East to Maryland; in wooded depressions with flowing water. Not in bogs. Frequently creates hummocks.

Microscope Tip Green cells are exposed more on the inner surface of the leaf in *Acutifolia* generally, though in this species the exposure is close to equal.

Name *fimbria* (L) = fringe, for the stem leaf.

subg. *Acutifolia*

S. fuscum sharing space with other low-nutrient specialists at the top of a hummock, Wonderland Bog, Mount Desert Island, ME; inset top left, actual size

Description Brown, with a small tightly packed capitulum; spreading branches short, <1 cm long with hanging branches 2× longer that clasp the stem; small branch leaves, ≈1 mm long, with inrolled margins, appearing very sharp-pointed, in 5 neat rows (5-ranked); stem is brown, especially in lower parts.

Fascicles 4–5 branches, with 2 spreading.

Stem Leaf About the same size as the branch leaves, ≈1 mm long, with rounded apex.

Similar Species The brown color, small size, neatly ranked leaves, and reliable hummock-top habitat provide a very straightforward ID. *S. flavicomans* is larger, and without the neatly ranked leaves. *S. pulchrum* in subg. *Cuspidata* forms lawns in bogs at water level. See comparison, page 42.

fuscum

Range and Habitat On hummock tops in ombrotrophic bogs throughout our range.

Name *fuscus* (L) = brown.

subg. *Acutifolia*

Top, *S. pulchrum* (subg. *Cuspidata*); middle, *S. fuscum* (subg. *Acutifolia*); bottom, *S. flavicomans* (subg. *Acutifolia*), 5×

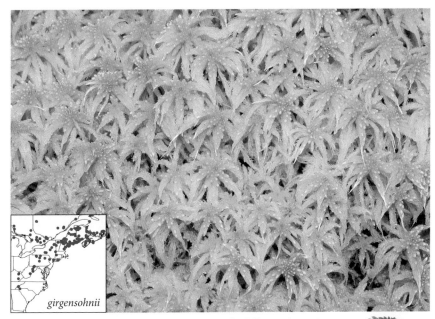

S. girgensohnii in a roadside seep, Nelson, NH; right, full-size plant. Note the single branch bud between rays of the stellate capitulum, and the quite horizontal spreading branches. Below right, top view of the apical bud.

Description Always green; large plants with stellate capitulum and a frequently prominent apical bud; with branches forming the capitulum rays often coalescing into a bearded look (above); single branch buds visible between rays (right, see arrow); branch leaves 1–1.5 mm long with rolled margins, appearing very sharp-pointed; stiff green stem snaps like celery.

Fascicles 3–5 branches, with 2 long spreading branches that leave the stem at 90 degrees and droop.

Stem Leaf Stem leaves approximately the same length as the branch leaves, with tip fringed and somewhat narrowed. See image (left), stained for better microscope detail.

Similar Species *S. fimbriatum* is usually smaller, but shares the color, the stellate capitulum with single branch bud, and the forest, or at least frequently shaded, habitat. Look for the more prominent apical bud on *S. fimbriatum*, and different stem leaves.

Range and Habitat Greenland to Alaska, south in the East to North Carolina and Tennessee; in boggy woods or seepy trailside, shade tolerant, not in open bogs or fens.

Name For Gustav Karl Girgensohn (1786–1872), Eastern European botanist. Common name: white-toothed moss.

Apical bud

subg. *Acutifolia*

S. rubellum forming an extended mat in full sun at Molly Bog, Stowe, VT; bottom right, actual size

Description Red, sometimes mottled, with a small capitulum, usually stellate with single hanging branches between rays, sometimes tightly packed and rounded; small branch leaves, ≈1 mm long, with rolled margins, appearing very sharp-pointed, in 5 neat rows (5-ranked); stem color red to greenish.

Fascicles 4–5 branches, with 2 spreading.

Stem Leaf ≈1 mm long, with rounded to slightly truncate apex.

Similar Species Will often be found on bottom of hummocks below *S. fuscum*, which it resembles in shape and size. The capitulum is often stellate versus pom pom–like for *S. capillifolium* and *S. subtile*. See *S. subtile* Similar Species for more comparisons.

S. rubellum, showing its neatly ranked leaves

Range and Habitat On hummock sides or lower, in ombrotrophic bogs and poor fens, often on floating mats; throughout our range.

Name *ruber* (L) = red.

rubellum

subg. *Acutifolia*

S. *russowii* in a mixed forest, Eagle Hill campus, Steuben, ME

Description Speckled red with a small to medium, flat-topped, stellate capitulum; single branch buds between rays; branch leaves with inrolled margins and a typical *Acutifolia* look, not 5-ranked; stem green or frequently with a with red tint.

Fascicles 3–4 branches, with 2 spreading.

Stem Leaf 1.2–1.5 mm long, lingulate, with a rounded, notched tip. See below.

Similar Species Many species show some pigment at the tips of spreading capitulum branches, often indicating the presence of antheridia; however, this spec shows spotty pigmentation in places other than branch tips. See S. *subtile* Similar Species.

Range and Habitat Throughout northeastern North America, south to Kentucky in the mountains; in shaded sites, often in a damp forest situation or drier wooded wetland margins.

Name Derivation Edmund Russow (1841–1897) was an Estonian botanis specializing in Sphagnaceae. He was one of the first researchers to see chromosomes, which he called *stäbchen*, or rods.

subg. *Acutifolia*

subtile

Above, on a relatively dry forest floor, Bald Head, Arrowsic, ME,
actual size; right and below, a shade population from Boothbay
Harbor, ME

Description Green in shade to very red in full sun, with a
small, tightly packed "pom pom" capitulum, which may be
stellate (see below), and if so, with single hanging branch
between rays; small branch leaves, ≈1 mm long, with rolled
margins, appearing very sharp-pointed, not 5-ranked; stem
green to yellowish green and stiff.

Fascicles 3–5 branches, with 2 spreading.

Stem Leaf ≈1 mm long, pointed.

Similar Species *S. capillifolium* and *S. rubellum* grow in
bogs, whereas this species never does. *S. capillifolium* has
stem leaves >1.5 mm long, compared with the stem leaves <1.2 mm long of this
species. *S. rubellum* usually has 5-ranked leaves and a stellate capitulum, while
S. capillifolium and *S. subtile* do not. *S. russowii*, also a forest floor species, is
more likely to be speckled and has rounded, notched stem leaves.

Range and Habitat From the Canadian Maritimes to Ontario, south in the
East to North Carolina; damp forest floor, not
in bogs.

Name *subtilis* (L) = slender. The English word
"subtle" makes more sense, in that the
differences between this species and *S.
capillifolium* and *S. rubellum* are subtle indeed.

subg. *Acutifolia*

Collected in Rockland Bog, a medium to rich fen, Rockland, ME; right, actual size, showing the single hanging branch between rays

Description These plants are distinctively reddish purple to green, often splotchy, as above; small to medium capitulum, densely branched, sometimes 5-parted, and if so, with 1 hanging branch between rays (see arrow, right); spreading branches long; branch leaves 1–1.5 mm long with inrolled margins, appearing very sharp-pointed, in 5 distinct rows when wet, somewhat recurved when dry; stem green or sometimes with a light reddish tint.

Fascicles 3–4 branches, with 2 spreading.

Stem Leaf Approximately the same size as the branch leaves, 1–1.5 mm long, with rounded apex. Note the photo bottom right showing (enhanced) stem leaves at the top of the stem with the capitulum popped off.

Similar Species The only rich-sited species in this subgenus, and one of the few rich-sited species in Sphagnaceae.

Range and Habitat A calciphile, found in mineral-rich habitats, such as rich fens or cedar swamps from the Canadian Arctic south to Florida.

Microscope Tip Note the green cells appropriately (for this subgenus) more exposed on the inner (concave) side of the leaf.

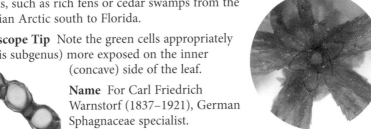

Name For Carl Friedrich Warnstorf (1837–1921), German Sphagnaceae specialist.

subg. Acutifolia   Ca

Subgenus *Cuspidata*

Size and Appearance From large to small, often elongate, slender, and weak-stemmed. McQueen (1990) suggested that these plants have the distinct look and feel of overcooked vegetables and a fishy odor.

Ecology Put your boots on. Many species in this group are found in very wet habitats from ombrotrophic bogs to sedge fens or open pools. *S. torreyanum* and some of the *recurvum* group can tolerate some shade, but the rest of the subgenus are typically found in the open.

Pigment Often yellowish or yellow-brown, rarely dark, never red, a characteristic that helps separate them from the frequently red-pigmented members of subg. *Acutifolia*, some of which are also found in wet situations.

Capitula Capitula usually are well developed, frequently stellate, and when clearly stellate, have paired emergent branch buds between rays. Subg. *Acutifolia* plants with stellate capitula always have single emergent branch buds between rays, so any stellate plants with 2 emergent branch buds showing are subg. *Cuspidata*.

Branch Leaves Frequently twisted and recurved when dry.

Stem Leaves Generally 1–1.5 mm, triangular with acute to blunt tip, divergent to divergent-pendent.

Name Cuspidate, meaning terminating in a point, refers to the often pointed stem leaves.

Microscope tip: In this subgenus, green cells are exposed more on the outer (convex) side of the leaf.

Key to Common Species in Subgenus *Cuspidata*

1. Capitulum stellate with paired branch buds between rays and branch leaves recurved when dry...*recurvum* group (see p. 49)
1. Capitulum not stellate 2

2. Dirty brown to greasy olive, in wet depressions ....*S. majus*
2. Brown to yellow-green, submerged, or floating...3

3. Branch leaves spirally 5-ranked, honey brown, forming lawns in a bog............
....*S. pulchrum*
3. Branch leaves not spirally 5-ranked, green to yellow-green, frequently at or below water level...4

4. Large spiky capitulum, strong stem ....*S. torreyanum*
4. Spiky capitulum, well developed or not, stem unable to support the capitulum (flops over)..*S. cuspidatum*

Recurvum Group

Four species in our range are commonly included in this complex, shown on pages 54 and 55. Until recently they were considered by many authors to be varieties of *S. recurvum*. They are similar and can be difficult to separate macroscopically or microscopically. I present a full page for *S. fallax* representing the group, and brief looks at the other members of the group: *S. angustifolium*,

angustifolium

S. flexuosum, and *S. recurvum*. You will need more technical resources to separate the members of the *S. recurvum* group definitively, but the photos and brief descriptions may get you started on the right track . . . or not.

Range and Habitat Any of these species may be found throughout the range of this book; species of the *S. recurvum* group are most common in early successional wetlands and can be found in a wide range of bogs and wooded fens at or slightly above water level.

fallax

flexuosum

recurvum

cuspidatum

Clockwise starting top left: typically submerged, actual size; capitulum close-up showing long, spiky outer branches; Bill Miller holding formless weak-stemmed aquatic plants; plants in drier site

Description Green to yellow with possible reddish tint at base of capitulum branches; capitulum sometimes poorly developed in inundated populations, more developed when stranded; branch leaves long, to 3 mm, appearing long-pointed by virtue of inrolled margins; generally tangled, flaccid, weak-stemmed plants that easily lose their heads; yellow-green stem.

When aquatic, as they frequently are, these plants are so weak-stemmed, tangled, and occasionally with such a poorly developed capitulum, that it isn't immediately clear that they are a *Sphagnum* sp.

Fascicles 4 branches, with 2 spreading, though very little difference between spreading and pendent branches—a characteristic more commonly found in subg. *Subsecundum*.

Stem Leaf ▷ ≈1 mm long with pointed apex, divergent to divergent-pendent.

Similar Species *S. torreyanum* also has long, inrolled leaves, and commonly grows submerged, but *S. torreyanum* generally has larger, spikier heads and is stiffer. In case of confusion, hold the plant by the stem and *S. torreyanum* will support its capitulum, and *S. cuspidatum* will collapse. See *S. majus* for comparison with that species.

Range and Habitat Throughout our range, Labrador to Ontario, south to Georgia; truly aquatic and likely to be found in pools in partial shade to full sun in bogs, poor fens, and roadside ditches. See map on page 49.

Name Cuspidate, a botanical term meaning abruptly pointed, is derived from the Latin *cusp*, meaning point, probably referring to the pointed stem leaves. The unfortunate common name is drowned kittens.

subg. *Cuspidata*

S. majus in a bog pool surrounded by *S. cuspidatum*, Aurora, ME; bottom right, actual size

Description Greasy looking, brown to olive to blackish; with a medium to large, spiky capitulum; capitulum branches occasionally with reddish tips; branch leaf 1.5–2.5 mm long with inrolled margins, a very pointed look; green stem.

Fascicles 4–5 branches, with 2 spreading.

Stem Leaf ▷ ≈1.5 mm long, triangular with blunt tip.

Similar Species This species is generally larger, spikier, and stiffer than *S. cuspidatum*, with which it frequently shares space, as shown above (see *S. cuspidatum* Similar Species). *S. torreyanum* shares the spiky look but is generally even more submerged, and it is green and shade tolerant.

Range and Habitat Labrador to Alaska, south in the East to Michigan and Virginia; grows in seasonally to continuously inundated depressions. Not shade tolerant.

Name The specific name indicates that this is one of the largest of the subgenus.

subg. *Cuspidata*

pulchrum

Wonderland Bog, Mount Desert Island, ME; above, actual size

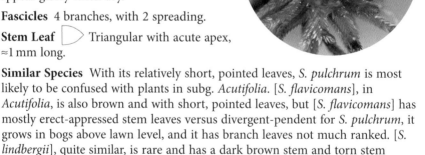

Description Orange-brown to yellow-brown, often with some green, as shown on the right; robust with densely branched capitulum; branches plump and short; branch leaves 5-ranked, ≈1.5 mm long, broad, but with such inrolled margins that they appear pointed; stem pale yellow to brown. Plants appear glossy when dry.

Fascicles 4 branches, with 2 spreading.

Stem Leaf ▷ Triangular with acute apex, ≈1 mm long.

Similar Species With its relatively short, pointed leaves, *S. pulchrum* is most likely to be confused with plants in subg. *Acutifolia*. [*S. flavicomans*], in *Acutifolia*, is also brown and with short, pointed leaves, but [*S. flavicomans*] has mostly erect-appressed stem leaves versus divergent-pendent for *S. pulchrum*, it grows in bogs above lawn level, and it has branch leaves not much ranked. [*S. lindbergii*], quite similar, is rare and has a dark brown stem and torn stem leaves. *S. fuscum*, another brown subg. *Acutifolia* species, is smaller with a tighter capitulum, and grows high on hummocks. See comparison, page 42.

Range and Habitat Forms floating lawns at water level in ombrotrophic bogs. Not shade tolerant. Common in the northern part of our range, where most ombrotrophic bogs are found.

Name *pulchra* (L) = beautiful. For good reason.

subg. *Cuspidata*

Above, floating in a shaded pool, Mount Desert Island, ME; right (actual size) and lower right, from a stream in Phippsburg, ME

Description Large, stiff, green plants with a round spiky capitulum, sometimes 5-parted, sometimes not. Plants can vary from small to elongate. Branch leaves not ranked, very long, inrolled, straight, ≈4.5–6 mm long; green stem.

Fascicles 4 branches, with 2 spreading.

Stem Leaf ▷ Triangular, ≈1.5 mm long, acute to blunt.

Similar Species *S. cuspidatum*, quite similar, is not as robust or spiky as this. See *S. cuspidatum* Similar Species for distinctions.

Range and Habitat Primarily in coastal areas from the Canadian Maritimes south to New York, though the range is confusing. It may be rare as far south as Florida and Louisiana. Sporadic and uncommon throughout our range; aquatic, shade tolerant, frequently in pools or in moving water with just part of the capitulum exposed.

torreyanum

Name For John Torrey (1796–1873), New York physician and botanist, founding president of the Torrey Botanical Club.

Note See page 24 for a look at a stranded plant with capsules.

subg. *Cuspidata*

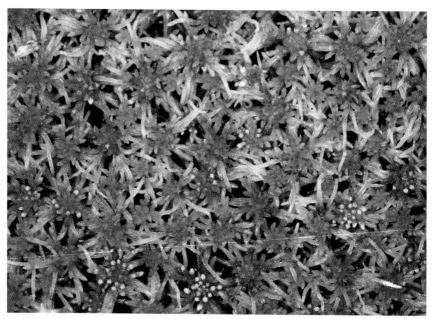

The dominant moss in an extensive shrub swamp, Woodland Bog, Baileyville, ME; above, actual size

Description Green to yellow-brown; medium-sized with dense, stellate capitulum and paired branch buds between rays (lower right, see arrow); outer capitulum branches long-tapering, and frequently twisted around the plant axis; branch leaves to 2 mm long, straight when moist, recurved when dry (right), not or only weakly 5-ranked; pale green to yellowish stem.

Fascicles 4–5 branches, with 2 spreading.

Stem Leaf ▷ Short, triangular, pointed, ≈1 mm long.

Recurved leaves

Similar Species See subg. *Cuspidata* page.

Range and Habitat The group is found throughout our range, from the Canadian Arctic south to Florida; frequently dominant at water level in bogs and poor fens, also in forested wetlands and ditches. *Recurvum* group is often found in early succession situations. See page 49 for distribution maps.

Name *fallax* (L) = deceptive. A most appropriate name for a member of this taxonomically troublesome group.

Paired branch buds

Continued on next page

Sphagnum angustifolium

Description Slender, soft, green to yellow-brown plants; stems frequently pink, providing the reddish tinge shown; spirally ranked branch leaves; long hanging branches that only loosely cover the stem; smallest capitulum of the group.

Name *angustus* (L) = slender, for the narrow leaves.

actual size

Sphagnum flexuosum

Description Pale green to yellow-brown; capitulum composed of tangled branchlets; with stems pale green to yellowish; branch leaves not 5-ranked; Richard Andrus (1980) described this as a "tight ball of yarn surrounded by a spiral nebula."

Name *flexuosus* (L) = winding or bending, for the outer capitulum branches.

actual size

Sphagnum recurvum

Description Large pale green to yellowish plants with a full, dense capitulum; pale stems; branch leaves straight, 1–1.5 mm long, 5-ranked. Largest plants of the group.

Name For the branch leaves that recurve when dry.

actual size

subg. *Cuspidata*

S. *wulfianum* in wet woodland, Belfast, ME; above, actual size

This, the only species in subg. *Polyclada,* is sometimes placed in subg. *Acutifolia.*

Description Large robust plants with a well-developed, densely packed hemispherical capitulum; branch leaves with recurved margins, ≈1 mm long, appearing pointed and neatly arranged in rows, leaf tips recurved when dry; stem stiff, dark brown to black.

Fascicles 6–12 branches, with 3–6 spreading (below right).

Stem Leaf Stem leaves are very small, 0.5–0.8 mm long, pointed or slightly rounded.

Similar Species As Roger Tory Peterson said of the bald eagle, this species is "all field mark." With its relatively dry, rich forest site; separate, not mat-forming plants; large capitulum; fascicles with more than 6 branches; and 5-ranked branch leaves, this is an easy ID. Too bad it's not more common.

Range and Habitat Primarily in the northern part of our range, from Pennsylvania north; a forest floor species, in cedar swamps or other rich forest sites.

Name The subgenus is named for the many branches per fascicle (*clade* (L) = branch). The species name is for Nils Gregers Ingvald Wulfsberg (1847–1888), a Norwegian doctor and plant collector who died at an early age when a steamer on which he was traveling sank.

Subgenus *Rigida*

This wet, seepy granite surface is being engulfed by a leading edge of the black liverwort, *Gymnocolea inflata*, followed by yellow-brown *Sphagnum compactum* and red *S. capillifolium*. Vascular plants complete the takeover; Welch and Dickey Mountain, NH.

These are very distinctive plants, characterized macroscopically by their almost tubular-looking, ovate, concave branch leaves, and their compact, mat-forming habit with colonies so dense that it is often difficult to discern individual plants. This subgenus comprises two species: *S. compactum* and [*S. strictum*] (more common in the Southeast coastal plain and Florida).

  S. compactum, like many sphagna, has antibiotic qualities and a great ability to hold moisture. This species and species of subg. *Sphagnum* were collected for use as surgical dressing as recently as World War I.

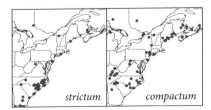

subg. *Rigida*

Typically dense *S. compactum* on seepy rock, Eagle Hill Campus, Steuben, ME; below right,

Description Light green to yellow-brown; medium to large plants with crowded capitula frequently not individually noticeable due to the tight, compact growth form; capitulum branches grow upward; large, concave, almost cucullate branch leaves to 3 mm long are usually somewhat spreading, to mildly squarrose; dark brown stem.

Fascicles 4–5 branches, with 2 spreading, short and fat.

Stem Leaf ▷ Small, ≈0.5 mm long, with erose tip.

Similar Species [*S. strictum*], rare in the north, favors sandy soil and is more common from the Pine Barrens of New Jersey south. It is very white, lacking much of the brown pigmentation in *S. compactum*, and with a white to light green to less frequently light brown stem. *S. pylaesii* sometimes shares the very crowded growth form and the seepy habitats, but is otherwise quite different.

Range and Habitat Throughout our range from northern Canada south to Louisiana, Alabama, and Georgia, though often replaced by [*S. strictum*] in the

South; seepy situations on wet sand or rock, in full sun to shade. Shade populations are more likely to have well-separated individual plants.

Microscope Tip The branch leaf drawings (left) show the slightly truncate tip frequently visible only with a microscope.

Name For the tight growth form.

subg. *Rigida*

Subgenus *Sphagnum*

Typical branch leaves for subg. *Sphagnum*: fat, hooded, and with some roughness at the tip. This is S. *papillosum*.

These are large robust plants, characterized macroscopically by their ovate, concave, hooded (cucullate) branch leaves. Other subgenera include plants with similar ovate leaves and inrolled sides (see subg. *Rigida*), but only subg. *Sphagnum* leaves have margins joined at the apex to form a complete hood. Stem leaves in this subgenus are all quite similar; lingulate, ≈1.5–2.0 mm long.

When the members of this group have their characteristic color and are growing where we expect them to grow, they can frequently be identified in the field; however, these species are often green and they don't always follow habitat instructions. So you may have to be satisfied with a subgenus determination, pending examination with a compound microscope.

Our Common Species

S. *centrale*: Usually pale green, never red, may infrequently have a light brown tint; found in cedar swamps or other rich, swampy, wooded sites.

S. *imbricatum*: A green to orange-brown bog and fen species; commonly divided into two species, S. *affine* and S. *austinii*, separable by microscopic characters.

S. *magellanicum*: Red coloration is common; frequently in or near ombrotrophic bogs in sun to partial shade.

S. *palustre*: Green in spring, frequently developing a golden-brown color; long spreading branches; on damp, acidic forest floor, not usually in bogs.

S. *papillosum*: Also commonly with brown pigment (see above); short, stubby, cigar-shaped branches in capitulum; usually in habitats with some enrichment such as minerotrophic bogs or poor to medium fens, occasionally in rich, shaded forests.

Microscope Tip Outer stem and branch cells of subg. *Sphagnum* have reinforcing diagonal fibrils not typically found in other subgenera.

subg. *Sphagnum*

In a treed wetland at base of Mt. Pisgah trail,
Winthrop, ME; top, approximately full size

Description Whitish green to light green,
robust with a well-developed capitulum; long
tapering outer capitulum branches; large,
cucullate branch leaves 1.5–2.5 mm long;
green to light brown stem.

Fascicles 4–6 branches, with 2 spreading.

Stem Leaf 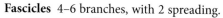 Widest above middle, rounded at tip with margin ragged
across the top and sometimes toothed down the sides, ≈1.5 mm long.

Similar Species *S. palustre* (often brown), and *S. magellanicum* (often red) have
similar capitulum shape and can be impossible to macroscopically separate
from *S. centrale* if green. *S. papillosum* (often brown), the other rich-sited
member of this subgenus, has shorter and stubbier outer capitulum branches.

Range and Habitat Throughout northeastern North America, south to
Pennsylvania, possibly farther south in the mountains; in damp, rich wooded
sites, most commonly in cedar swamps. Shade tolerant, in pH up to 7.8
(McQueen, 1990).

Microscope Tip Note in the photo below how the green
cells are open to the environment on both sides of the leaf
through channels between
the hyaline cells.

Name *centralis* (L) =
center, for the location of
the green cells in relation
to the hyaline cells.

centrale

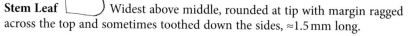

S. austinii (identified from microscopic characteristics) collected in Jonesport, ME, from a previously harvested peat bog; inset top left, actual size

Description Green to orange-brown, large, with a well-developed capitulum, forming tightly packed colonies, often with fugitive branches, as shown above; branch leaves large, hooded, 1.5–2.0 mm long; stem brown to dark brown.

Fascicles 3–4 branches, with 2 spreading.

Stem Leaf Short rectangular, 0.8–1.1 mm long.

Similar Species When in a bog, the orange-brown color, tightly packed growth, and fugitive branches provide for reasonably straightforward identification. In other habitats, or when green, an identification would require microscope work. See Note below.

Range and Habitat From the Canadian Maritimes south to Florida; appears largely confined to locations within 200 miles of the coast; found in bogs in the northern part of the range, treed swamps and water margins in the southern part of the range.

Name *imbricatum* is for the tightly packed (imbricate) leaves; *austinii* is for Coe Austin, collector of the type specimen, the specimen used to initially describe the species, in New Jersey in 1862; *affine* (L) = related, for the close connection to *S. cymbifolium*, an early name for *S. palustre*.

Note *S. affine* and *S. austinii* were once considered one species, *S. imbricatum*. The differences are microscopic, and a matter of much discussion among experts. I lump *S. affine* and *S. austinii* together as *S. imbricatum* s.l. (*sensu lato*, meaning in the wider sense) for convenience.

subg. *Sphagnum*

Red *S. magellanicum* with brown *S. fuscum* (possibly
[*S. flavicomans*]); right and below show a range of
coloration; top, actual size

Description Very red to green, but almost always
with some red coloration, especially in sun; robust
with a well-developed capitulum; large, cucullate
branch leaves 1.5–2.0 mm long; red stem.

Fascicles 4–5 branches, with 2 spreading.

Stem Leaf  About as long as branch leaves,
rectangular to wider above middle, with rounded,
sometimes fringed tip.

Similar Species This is the only red member of the
subgenus. When green, it may be impossible to
macroscopically separate from other species in the
subgenus.

Range and Habitat Widespread throughout our
range, and throughout much of the world; a mat and
hummock former in nonaquatic portions of acidic
bogs. Occasionally in shaded treed swamps, where it is
more likely to be green.

magellanicum

Microscope Tip Note how the
green cells are enclosed by the
hyaline cells.

Name The type specimen for this
species was collected in the Straits
of Magellan.

subg. *Sphagnum*

Above, a typical plant with brown tint; right, a green example
with brown stem; inset, actual size

Description Golden-brown to green (particularly in
spring), but reliably with some brown at least in the
lower stem; robust with a well-developed capitulum;
long branches; large, cucullate branch leaves
2.0–2.5 mm long can be somewhat spreading, as
shown.

Fascicles 4–5 branches, with 2 spreading.

Stem Leaf  Almost as long as branch leaves
(1.5–2.0 mm), rectangular, with rounded, sometimes
fringed tip.

Similar Species *S. papillosum*, the other brown subg.
Sphagnum species, has shorter branches and different
ecology. When green, *S. palustre* can easily be confused
with green forms of *S. magellanicum* and *S. centrale*, and
microscope sections will be necessary for identification.
When green and squarrose, *S. palustre* can be confused
with *S. squarrosum*.

Range and Habitat Scattered throughout our range in
shaded, acidic, damp forest floor sites.

Microscope Tip Note green cells are more exposed on the
inside (concave, or lower side in this illustration) of the leaf.

Name *palustre* (L) = marshy. A bit of a misnomer for this
species of damp, shaded, frequently forested situations.

palustre

subg. *Sphagnum*

Above, *S. papillosum* in a bog hummock; below right, actual size, illustrating how stringy the plants can get when growing in a wet situation

Description Usually golden brown to occasionally more green, but reliably with some brown at least in the stem; robust with a well-developed, medium-sized capitulum; short cigar-shaped branches in the capitulum; large cucullate branch leaves to 2.0 mm long; dark brown to black stem.

Fascicles 4 branches, with 2 spreading.

Stem Leaf Almost as long as branch leaves, ≈1.5 mm long, 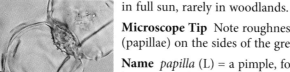gular, with rounded, fringed apex.

Similar Species *S. palustre*, the other brown subg. *Sphagnum* species, has longer branches and different ecology. *S. imbricatum* is similar and found in similar habitats, but it grows more tightly packed, with "escaping" capitulum branches. Microscope work may be required to separate the two.

Range and Habitat Scattered throughout our range in mats and lower parts of hummocks in open, poor to medium fens, usually in full sun, rarely in woodlands.

Microscope Tip Note roughness (papillae) on the sides of the green cells.

Name *papilla* (L) = a pimple, for the microscopic bumps (papillae) on the chlorophyll cells.

Subgenus *Squarrosa*

This subgenus is all about the leaves. Many species in other *Sphagnum* genera have leaves that are squarrose from time to time, usually when dry; however, in subg. *Squarrosa*, the leaves are squarrose whether wet or dry, and the resulting fuzzy look makes a viewer feel as though a trip to the eye doctor is in order.

Two species occur in the Northeast, *S. squarrosum*, illustrated here, and [*S. teres*], rare, but widespread across northern North America. From New England and upstate New York north, the ranges of the two species overlap. [*S. teres*] may be very squarrose in the upper part of the plant, less so in lower parts. Its stem leaves are similar in shape and size to those of *S. squarrosum*, but [*S. teres*] branch leaves are smaller, <1.5 mm long versus 2–3 mm long for *S. squarrosum*. Also, [*S. teres*], a calciphile, grows in cushions in rich fens while *S. squarrosum* grows as individual plants, and is a species of damp, shaded forests.

Seriously squarrose!

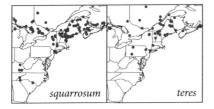

squarrosum   teres

subg. *Squarrosa*

These plants were in a damp conifer forest near West Branch Pond Camps, Shawtown Township, ME; lower right, actual size.

Description Large, bright yellow-green plants with a full, shaggy capitulum and prominent apical bud (below left); large, squarrose, inrolled, branch leaves to 3 mm long; green stem.

Fascicles 4–5 branches, with 2 spreading.

Stem Leaf Large, 1.5–2.0 mm long, with tip occasionally torn.

Similar Species This is the only subgenus with squarrose leaves wet or dry.

Range and Habitat From Newfoundland and Labrador south to New York and Pennsylvania, farther south in the mountains in wet coniferous forests, or other rich, shaded situations.

Name Describing the leaves.

subg. *Squarrosa*

Subgenus *Subsecunda*

Some Caveats

In 1958, A. LeRoy Andrews, noted American bryologist, again attempted, for at least the second time, to straighten out the *S. subsecundum* complex. He referred to the taxonomic situation left by his predecessor, Carl Warnstorf (German botanist, 1837–1921) as a *"monstrum horrendum, informe, ingens, cui lumen ademptum,"* which means something like "an immense, horrible, amorphous, blind monster." It's still a pretty confusing group.

Crum and Anderson (1981) refused to give specific rank to *S. subsecundum* varieties, saying, "we prefer keeping aquatics of variable nature at a low taxonomic level." More recently, Jon and Blanka Shaw, researchers at Duke University, have revised this group based on DNA analysis, providing some much-needed clarity; however, the first division in their key is whether the stem cortex has one or two (plus) layers of clear cells on the outer surface, and another important split is based on the number of cells at the tip of a stem leaf that appear divided by fibrils. It's just too technical for this book, though we'll give you an idea of what they're talking about in one of the microscope tips below.

These plants are aquatic, for the most part. And as with many aquatic plants, being freed from at least some of the constraints of gravity and water transport, they show a high degree of morphological diversity.

The Odd Ducks

S. pylaesii doesn't greatly resemble the rest of the section ecologically or visually, and has only recently been included in this subgenus by virtue of DNA analysis. Please see page 69 for this unusual species.

[*S. platyphyllum*] is also a morphological outlier. It is rare, with a poorly developed capitulum, sparse branches of only 2 or 3 barely differentiated branches per fascicle, and stem leaves as large as branch leaves.

S. subsecundum s.l.

The rest of the species in the group can be included in *S. subsecundum* s.l. (for *sensu lato*, or "in the broad sense"). The species (or varieties if you prefer) are all aquatic to seasonally inundated, with a light green to yellow-brown, usually well-developed capitulum, often with upturned "bull horn" branches. Dried specimens frequently show a bluish tint on lower parts. The branch leaves are inrolled for most of their length and curve in an organized way.

Microscope Tip In this subgenus, the green chlorophyll cells are exposed equally on the upper and lower surface of the branch leaf.

Most members of the group will all be described pretty well by the page for *S. subsecundum* s.l. Following are the species (varieties for some authors) included within the broad tent of *S. subsecundum* s.l. and might be separately identified with technical resources.

Species Included within *S. subsecundum* s.l.

S. subsecundum s.l.

[*S. contortum*]: Weak-stemmed, in rich sites, intolerant of shade, stem leaf <0.8 mm long. This is the species used to illustrate the *S. subsecundum* s.l. account (p. 70)

[*S. lescurii*]: Seasonally flooded to inundated, stem leaf >1.0 mm, shade tolerant, a more southern species, not found north of Massachusetts. Not visually separate from *S. missouricum*.

S. missouricum: Seasonally flooded situations, stem leaf >1.0 mm, throughout our range. Not visually separate from [*S. lescurii*] (p. 71).

Microscope Tip *S. subsecundum* s.l., [*S. lescurii*], and *S. missouricum* have a single layer of clear outer stem cells, as shown here. [*S. contortum*] and [*S. platyphyllum*] have a double to quadruple layer.

On seepy rock, Eagle Hill campus, Steuben, ME, showing color variation; inset top left, actual size

This species will be included in many references within subg. *Hemitheca*.

Description Slender, with small capitulum; stem leaves ≈ 2.0 mm long, much larger than similarly shaped branch leaves, ≈ 1.0 mm long; single branches (occasionally 2) per fascicle. Much color variation, as shown. On seepy sites, the plants coalesce and appear as an undivided mass.

Fascicles Usually single branches.

Stem Leaf Very large, to as much as 2.5 mm long.

Similar Species This species appears to have a unique suite of characteristics, though *S. compactum* also favors succession sites and shares the habit of coalescing in such a way that individual plants may not be obvious.

pylaesii

Range and Habitat Northern New England and the Canadian Maritimes; on wet rock succession sites, and in bog depressions.

Name Bachelot de la Pylaie (1786–1856), a French explorer and scientist, was one of the first Europeans to collect in Newfoundland and on St. Pierre and Miquelon islands.

subg. *Subsecunda*

In a shaded, boggy, seasonally wet roadside, Arrowsic, ME; right, actual size

Description Green to yellow-brown; slender, small to medium plants with a well-developed, not 5-parted, capitulum; branch leaves to 2 mm long, frequently curved upward (see below right and next page); stem green above to splotchy brown below.

Fascicles 4–5 branches of similar length, with 2 spreading.

Stem Leaf Stem leaves small, triangular, 0.5–0.8 mm long, pointed, divergent, and mostly descending (below left).

Similar Species See subg. *Subsecunda* page.

Range and Habitat Throughout our range, from Pennsylvania north in wet to seasonally inundated treed wetlands, tolerant of shade, or not.

Name *secundus* (L) = to follow, for the aligned branch leaves.

subg. *Subsecunda*

S. missouricum (included in *S. subsecundum* s.l.) showing the yellow-brown color common in less shaded situations, and nice "bull horns" in the capitulum. Collected under shrubs in partial shade along a pond margin in Brunswick, ME.

subg. *Subsecunda*

Acrocarpous Mosses

Almost all acrocarps have a noticeable single costa, have erect, rarely branched to occasionally forked stems, grow in cushions or tufts, and produce sporophytes from the tip of a major stem or branch.

The family Fissidentaceae is included here, though it includes some species in which sporophytes originate from well below the stem tip. Fortunately, this family is easy to recognize.

Philonotis fontana (Bartramiaceae) is an example of a potentially confusing acrocarp by virtue of its branching at the base of terminally produced antheridial heads. See photos on page 107.

Also, in some species, a stem will produce an innovation below the tip and continue to grow after producing a sporophyte, resulting in the appearance of the sporophyte originating from the stem below the tip, mimicking a pleurocarp. It is hoped that some teasing apart will show whether it is a little-branched and tuft-forming acrocarp or a branched and mat-forming pleurocarp.

Five families in this section include special treatments: Polytrichaceae, Dicranaceae, Fissidentaceae, Grimmiaceae, and Mniaceae.

SPECIES INCLUDED

The order followed here, alphabetically by family, reflects the order of the species accounts. Species in [brackets] are mentioned in keys, but not illustrated.

Andreaeaceae (p. 101)
Andreaea rothii
A. rupestris

Aulacomniaceae (p. 103)
Aulacomnium androgynum
Aulacomnium heterostichum
Aulacomnium palustre

Bartramiaceae (p. 106)
Bartramia pomiformis
Philonotis fontana
[*Plagiopus oederianus*]

Bryaceae (p. 108)
Bryum argenteum
Bryum pseudotriquetrum
Leptobryum pyriforme
Pohlia annotina
Pohlia bulbifera
Pohlia cruda
Pohlia nutans
Pohlia wahlenbergii
Rhodobryum ontariense
Rosulabryum capillare

Buxbaumiaceae (p. 117)
Buxbaumia aphylla
Diphyscium foliosum

Dicranaceae (p. 119)
Dicranella heteromalla
[*Dicranella varia*]
Dicranum flagellare
Dicranum fulvum
Dicranum majus
Dicranum polysetum
Dicranum montanum
Dicranum ontariense
Dicranum scoparium
Dicranum viride
Oncophorus wahlenbergii
Paraleucobryum longifolium
Trematodon ambiguus

Ditrichaceae (p. 133)
Ceratodon purpureus
[*Distichium capillaceum*]
[*Ditrichum flexicaule*]
[*Ditrichum lineare*]
Ditrichum pallidum
[*Saelania glaucescens*]

Encalyptaceae (p. 136)
Encalypta procera

Fissidentaceae (p. 137)
Fissidens adianthoides
Fissidens bryoides
Fissidens bushii
Fissidens dubius
Fissidens fontanus
Fissidens grandifrons
Fissidens osmundioides

Funariaceae (p. 140)
Funaria hygrometrica
Physcomitrium pyriforme

Grimmiaceae (p. 143)
Grimmia muehlenbeckii
Racomitrium aciculare
Racomitrium heterostichum group
Racomitrium lanuginosum
Schistidium apocarpum group
Schistidium maritimum

Leucobryaceae (p. 154)
Leucobryum albidum
Leucobryum glaucum

Mniaceae (p. 155)
Mnium hornum
Mnium spinulosum
[*Mnium stellare*]
Plagiomnium ciliare
Plagiomnium cuspidatum
Pseudobryum cinclidioides
Rhizomnium appalachianum
Rhizomnium punctatum

Orthotrichaceae (p. 164)
Orthotrichum spp.
Orthotrichum anomalum
Ulota coarctata
Ulota crispa
Ulota hutchinsiae

Polytrichaceae (p. 170)
Atrichum angustatum
Atrichum crispum
Atrichum undulatum (s.l.)
Pogonatum dentatum
Pogonatum pensilvanicum
Pogonatum urnigerum
Polytrichastrum alpinum

Polytrichastrum ohioense
Polytrichastrum pallidisetum
Polytrichum commune
Polytrichum juniperinum
Polytrichum piliferum
Polytrichum strictum
Pottiaceae (p. 185)
  Barbula unguiculata
  [*Barbula convoluta*]
  [*Bryoerythrophyllum*
    recurvirostrum]
  [*Syntrichia ruralis*]
  [*Tortella fragilis*]
  Tortella humilis

Tortella tortuosa
Tortula truncata
Weissia controversa
Schistostegaceae (p. 190)
  Schistostega pennata
Seligeriaceae (p. 192)
  Blindia acuta
Splachnaceae (p. 193)
  Splachnum pensylvanicum
Tetraphidaceae (p. 194)
  Tetraphis pellucida
Timmiaceae
  [*Timmia megapolitana*]

DICHOTOMOUS KEY TO GENERA AND SPECIES OF ACROCARPOUS MOSSES

1. Plants growing on dung or animal remains, often in bogs; capsules with swollen or expanded lower section.....................................*Splachnum* (p. 193)
1. Plants other than above 2

2. Plants erect, small, stems <2 cm high, brown to reddish brown or black, leaves with or without costa (midrib); capsules without teeth, opening most of their length by 4 longitudinal slits when dry (see right); on noncalcareous rock in full to partial sun....................................
.... *Andreaea* (p. 101)

Andreaea

A. Leaves ovate to ovate-lanceolate, ecostate (without midrib); common, on exposed acidic boulders and cliffs ....*A. rupestris* (p. 101)
A. Leaves narrowly lanceolate, costate (with midrib); less common, on shaded acidic boulders and cliffs .... *A. rothii* (p. 101)

2. Plants otherwise; capsules peristomate (with teeth) or eperistomate (without teeth), opening by an operculum (lid) 3

3. Plants apparently lacking leaves; capsules large, 5–7 mm long, and prominent, with a somewhat flattened upper surface on long warty setae; a rare and unusual moss in woods, soil under shrubs, clay banks, and on stumps and logs...*Buxbaumia aphylla* (p. 117)
3. Plants with conspicuous leaves; capsules usually smaller................................4

4. Plants with leaves in 2 distinct rows (distichous)*...5
4. Plants with leaves in more than 2 distinct rows*...7

*Several genera have leaves that are complanate but not distichous; that is, they appear flattened, but their leaves are actually inserted around the stem and twisted into one plane. Examples are *Neckera*, *Homalia*, *Plagiomnium*, and *Plagiothecium*.

5. Leaves narrow, <1 mm wide, lacking a flap fused onto the upper surface; on rock, sometimes calcareous, primarily in cliff crevices, occasionally on soil or humus over rock .... [*Distichium capillaceum*]
5. Leaves broad, more than 1 mm wide, with or without flap 6

6. Leaves with a flap fused onto the upper surface to form a large sheath; on damp shady humic soil, rotten logs, and seepy cliffs...
.... *Fissidens* spp. (p. 137)
6. Leaves without a flap fused onto the upper surface; plants with a reflective (appears to glow) protonema; on mineral soil, under tree tip-ups, in cave mouths or other dark habitats .... *Schistostega pennata* (p. 190)

7. Plants with a large ovoid capsule, lacking an obvious seta and immersed among bristle-tipped leaves; on bare trailside or streamside banks.................
.... *Diphyscium foliosum* (p. 118)
7. Plants usually with a small capsule and a prominent seta, or if lacking a seta, capsule not ovoid and surrounded by bristle-tipped leaves.....................8

8. Plants greenish white, usually 3–6 cm high, in large, rounded, dense cushions on the ground; leaves tightly packed, subtubulose; on moist to dry soil or humus in woods, frequently along trails ....
.... *Leucobryum glaucum* (p. 154)
8. Plants not greenish white, or if so, not in dense cushions and without subtubulose leaves..9

9. Gemma cups (formed by apical leaves) frequently present at the tips of gemmiferous shoots; capsules cylindrical with 4 large peristome teeth; common, usually on rotting coniferous wood, occasionally on rock ledges....
.... *Tetraphis pellucida* (p. 194)
9. Gemma cups lacking; peristome teeth 16 or more; on various substrates.......
.... 10

10. Leaves with lamellae on upper surface, giving them an opaque look. Covering the genera *Atrichum, Pogonatum, Polytrichum*, and *Polytrichastrum*..see **Polytrichaceae** (p. 170)
10. Leaves lacking lamellae 11

11. Leaves narrow, often subulate, 10–20 or more times longer than leaf width near middle; costa sometimes covers most of leaf near middle.................... 12
11. Leaves broad, lanceolate, ovate, obovate, or oblong, mostly <10× as long as broad; costa often covering only a small portion of leaf near middle.......... 22

12. Leaves squarrose, with an enlarged base clasping the stem; capsules horizontal, strumose; on rotten logs, stumps, and tree bases............................
....*Oncophorus wahlenbergii* (p. 130)
12. Leaves not squarrose, or if squarrose, without enlarged clasping base; capsules erect to pendulous, struma lacking or indistinct; on a variety of substrates.. 13

13. Alar cells noticeably differentiated, often inflated and red to orange........... 14
13. Alar cells not noticeably differentiated... 16
> See discussion on differentiated alar cells on page 17.

14. Leaves short, ≈3 mm long, entire; stems short, ≈2 cm high, red; capsules rare, short, pyriform; on wet boulders, cliff faces, and ledges, frequently near waterfalls ....*Blindia acuta* (p. 192)
14. Leaves long, mostly >3 mm, serrate to serrulate on margins and back of costa (microscopic); stems long, nearly always >2 cm high, usually green or brown; capsules common, long, cylindrical; on rock or various substrates.....
.... 15

15. Costa broad, occupying ≈2/3 of leaf at base and nearly all the leaf near the middle; capsules straight, erect, usually on rock or soil over rock...................
....*Paraleucobryum longifolium* (p. 131)
15. Costa narrow, occupying ≈1/3 of leaf at base and usually less above; capsules straight to arcuate, erect or inclined; on various substrates...............
....*Dicranum*
> See **Dicranum** key (p. 120).

16. Capsules distinctly narrowed at a neck that is nearly as long as the urn (especially noticeable when dry); leaves entire... 17
16. Capsules not narrowed at neck; leaves entire or serrate............................... 18

17. Capsules cylindrical, indistinctly strumose, erect to somewhat inclined; on soil in open, disturbed sites..............................*Trematodon ambiguus* (p. 132)
17. Capsules pyriform, not strumose, horizontal to pendulous; on soil, rock, or rotten wood, often in burned-over or disturbed habitats, common in greenhouses ....*Leptobryum pyriforme* (p. 110)

18. Leaves with recurved, serrate margins; capsules globose or nearly so......... 19
18. Leaves with plane or incurved margins, entire or indistinctly serrate near tip; capsules cylindrical.. 20

19. Leaves light green, mostly 4 mm long or more, crisped and contorted when dry, base sheathing stem; common, on soil or humus over noncalcareous boulders, on cliff ledges or in crevices ....*Bartramia pomiformis* (p. 106)
19. Leaves dark green to yellowish green, seldom reaching 4 mm in length, weakly contorted when dry, base not sheathing stem; on calcareous boulders and cliffs, especially along streams.................*[Plagiopus oederianus]*

20. Leaves with a V-shaped region of hyaline cells at base, dorsal surface with a dull lamina and a distinctly shiny costa, leaves sometimes with broken tips and missing portions.. *Tortella*

Tortella

A. Leaves with upper portion often broken off; on calcareous soil or rock, often on cliff ledges or in crevices...................................*[T. fragilis]* (p. 186)
A. Leaves usually intact; on calcareous soil or rock, often on cliff ledges or in crevices.. B

    B. Leaves to 6 mm long, tapered to a long point..............*T. tortuosa* (p. 187)
    B. Leaves to 4 mm long, with an obtuse mucronate tip (see also couplet 29)
    ....*T. humilis* (p. 186)

20. Leaves lacking V-shaped region of cells and differentiated lamina and costa, seldom broken.. 21

21. Peristome teeth filiform; capsules not contracted under mouth when dry......
    ....*Ditrichum*

Ditrichum

    A. Stems usually tomentose, often >1 cm high; leaves 2–7 mm long; sporophytes rarely produced; on calcareous rock or soil over rock, especially on bluffs, cliff shelves, or in cliff crevices (rare) ....
    .... [*D. flexicaule*] (p. 135)
    A. Stems not tomentose, seldom reaching 1 cm high; sporophytes common, setae yellow; on acidic clay, sand, or gravelly soil banks in wooded clearings, along trails, roads, or in other disturbed habitats .... B

    B. Leaves 2–7 mm long..*D. pallidum* (p. 135)
    B. Leaves 1–2 mm long..[*D. lineare*]

21. Peristome teeth lanceolate; capsules strongly to slightly asymmetrically contracted just below the mouth when dry..*Dicranella*

Dicranella

    A. Capsules strongly contracted under a portion of mouth when dry; setae yellow to brown ....*D. heteromalla* (p. 122)
    A. Capsules scarcely contracted under mouth when dry; setae red................
    .... [*D. varia*]

22. Alar cells strongly differentiated, inflated, often red or orange.......................
    ....see **Dicranum** key (p. 120)
22. Alar cells not noticeably differentiated... 23
              See discussion on differentiated alar cells on page 17.

23. Leaves entire or nearly so at apex 24
23. Leaves serrate or toothed at apex... 39

24. Leaves with a differentiated marginal border that is often thicker than the lamina... 25
24. Leaves lacking differentiated border... 26

25. Leaves lanceolate to ovate or oblong-lanceolate, seldom >1 mm wide, acute to acuminate, often with a long-excurrent costa ....*Bryum*

Bryum/Rosulabryum

    A. Plants silvery white; stems short, 0.4–1.0 cm high; leaves nondecurrent, costae ending below apices; on predominantly dry soil in disturbed habitats, especially in cracks of sidewalks, along paths, roads, and railroads ....*B. argenteum* (p. 108)

A. Plants not silvery white, i.e., not lacking chlorophyll in leaf tips; costae percurrent to excurrent .... B

B. Plants green or sometimes brownish or reddish; stems long, 2–6 cm high; leaves long-decurrent, costae percurrent to short excurrent; on wet, often sandy soil or humus beside roads, streams, or lakes, sometimes on wet boulders and rock ledges that are frequently calcareous, and occasionally on decayed wood in swamps .... *B. pseudotriquetrum* (p. 109)

B. Plants green or sometimes brownish or reddish; stems short, <1 cm high; leaves 1–2 mm long with excurrent, frequently red costa; on tree bases, soil, often soil over rock, frequently in calcareous situations .... *Rosulabryum capillare* (p. 116)

25. Leaves obovate to rounded ovate, usually >1 mm wide, costa ending below apex or in a short blunt mucro .... *Rhizomnium,* see **Mniaceae** (p. 155)

26. Plants on tree trunks, often in small rounded tufts, rarely on fallen trees or logs 27

26. Plants on soil, rock, humus, sometimes at bases of trees and over rotting logs, but not on tree trunks 28

27. Leaves obtuse, margins revolute; capsules immersed; on deciduous tree trunks, especially elm, maple, poplar, and willow, rarely on conifer trunks.... .... *Orthotrichum obtusifolium* (p. 164)

27. Leaves acute, margins recurved; capsules exserted, on deciduous tree trunks or rock .... *Ulota*

Ulota

A. Plants with leaves strongly crisped when dry; occurring on tree trunks and limbs .... *U. crispa* (p. 168)

A. Plants with leaves straight, curved, or twisted, but not crisped when dry; occurring on tree trunks and limbs, or rock.. B

B. Plants on tree trunks or limbs; capsules pyriform with a small puckered mouth .... *U. coarctata* (p. 168)

B. Plants on rock, capsules cylindrical, without a puckered mouth; usually on noncalcareous rock in forests.............................. *U. hutchinsiae* (p. 168)

28. Leaves obtuse (leaves near the stem tips) to broadly acute............................ 29
28. Leaves acute to acuminate.. 30

29. Capsules exserted above leaves on a long seta; peristome teeth filamentous, long and twisted; on calcareous soil in open, disturbed habitats, such as roadsides, gravel pits, fields.............. *Barbula* [*convoluta*]/*unguiculata* (p. 185)

See also *Tortella humilis* at couplet 20.

29. Capsules immersed on a short seta; peristome teeth lanceolate, short and not twisted; on rock or beside streams or beside ocean...................................... .... *Schistidium,* see **Grimmiaceae** (p. 143)

30. Plants nearly julaceous, mostly silvery white because of an absence of chlorophyll in cells... *Bryum argenteum* (p. 108)
30. Plants neither julaceous nor silvery white 31

31. Leaf margins inrolled when dry; on soil, soil over rock, or in rock crevices in usually calcareous, disturbed, exposed habitats, especially roadside banks and fields.. *Weissia controversa* (p. 189)
31. Leaf margins recurved or plane when dry.. 32

32. Capsules inclined to pendulous, sulcate, ribbed, or wrinkled when dry..... 33
32. Capsules erect, smooth, or ribbed when dry 35

33. Leafy plants bulbiform; setae flexuose; capsules pyriform (pear-shaped) and almost always present; a common weed in disturbed habitats, especially on burned wood and on soil in roadside ditches and greenhouses.......................
.... *Funaria hygrometrica* (p. 140)
33. Leafy plants not bulbiform; setae straight or vertically twisted; capsules cylindrical 34

34. Leaves green, with some red or purple coloration near the base; pseudopodia lacking; capsules and setae red or purple when fresh, brown at maturity; an extremely common, weedy species found in disturbed habitats on soil, rock, wood, and humus ....*Ceratodon purpureus* (p. 133)
34. Leaves yellowish green, without red or purple color; pseudopodia often present with gemmae attached; capsules and setae yellow or brown, sometimes reddish brown .... *Aulacomnium*

Aulacomnium

A. Plants small to medium-sized with lanceolate leaves; gemmae frequently present on pseudopodia .... B
A. Plants somewhat complanate, medium-sized, with stems to 4 cm long; leaves ovate with tips blunt and toothed; gemmae not present, capsules frequent; in rich woods ....*A. heterostichum* (p. 104)

B. Medium-sized to robust plants, stems 3–9 cm high; pseudopodia present on some plants, bearing clusters of tiny leaflike gemmae at apices, often with only naked pseudopodia after gemmae have fallen; on soil or humus, sometimes on rotting logs, often in bogs and swamps, at lake margins, beside streams, or in other wet habitats........*A. palustre* (p. 105)
B. Small plants, stems 1–4 cm high; pseudopodia present on some plants, bearing round clusters of gemmae at apices; generally in dry habitats, on soil or humus in coniferous woodlands, often over rock, sometimes on decaying wood ....*A. androgynum* (p. 103)

35. Leaves mucronate, the mucro often yellowish; often calcareous habitat 36
35. Leaves blunt to acute; calcareous or not 38

36. Leaf margins recurved; on calcareous soil or soil over rock, in open, usually disturbed habitats, especially roadsides and woodland trails...........................
.... *Barbula unguiculata* (p. 185)
36. Leaf margins plane 37

37. Capsules obovate, opercula oblique, calyptrae cucullate, peristome absent; in open weedy habitats such as roadsides, lawns, gardens, fields, and pastures, on bare, often calcareous soil .... *Tortula truncata* (p. 188)

37. Capsules cylindrical, opercula oblique, calyptrae mitrate, peristome present (unfortunately, capsules rarely produced); in woodlands on rock ledges or in crevices of calcareous bluffs or cliffs, sometimes on soil over boulders....... .... *Encalypta procera/ciliata* (p. 136)

38. Leaf apices sometimes blunt; capsules with 16 ribs, 8 long ones alternating with 8 short ones; occurring on calcareous rock.. .... *Orthotrichum anomalum* (p. 167)

38. Leaf apices acute; capsules with 8 ribs of about the same length, occurring on noncalcareous rock .... *Ulota hutchinsiae* (p. 168)

39. Capsules globose, subglobose, or pyriform 40

39. Capsules cylindrical to ovoid... 41

40. Plants large, 3–8 cm high; capsules inclined, ribbed when dry, peristome present; perennial, on soil, often over rock, in wet places, especially in roadside ditches and along streams .... *Philonotis fontana* (p. 107)

40. Plants small, <3 cm high; capsules erect, smooth or wrinkled at base when dry, peristome lacking; an annual, occurring in spring on moist, bare, exposed soil, in disturbed habitats, such as streambanks, roadside ditches, lawns, pastures, and meadows .... *Physcomitrium pyriforme* (p. 142)

41. Leaves with a differentiated border, the border of a lighter color and often thicker than the lamina.. 42

41. Leaves lacking differentiated border 45

42. Leaf border hyaline with irregular teeth in the upper part................................. .... *Racomitrium lanuginosum* (see couplet 49), ...and see **Grimmiaceae** (p. 143)

42. Leaf border not hyaline.. 43

43. Leaves crowded at stem tips, forming rosettes; plants connected by subterranean stems; on humus, rotting logs, bases of trees, and soil, often over limestone or in rich forest .... *Rhodobryum ontariense* (p. 115)

43. Leaves not crowded into rosettes and plants not connected by subterranean stems 44

44. Leaf borders with teeth in pairs (can be challenging to see with a hand lens); costa red, especially near base of leaves; stoloniferous shoots absent .... *Mnium*, see **Mniaceae** (p. 155)

44. Leaf borders with single teeth; costa yellow or green; stoloniferous shoots present .... *Plagiomnium*, see **Mniaceae** (p. 155)

45. Plants complanate (flattened); leaves oblong with teeth in upper part............ .... *Aulacomnium heterostichum* (in part) (p. 104)

45. Plants not as above 46

46. Leaves broad, often >3 mm wide; on soil or humus or in wet depressions in woodlands, sometimes on boulders or exposed tree roots. ....
....*Pseudobryum cinclidioides* (p. 161)

> *Rhizomnium* spp. may also key here; see **Mniaceae** (p. 155)

46. Leaves narrow, <3 mm wide 47

47. Leaves ending in a long hyaline point or awn; capsules cylindrical, erect, immersed or exserted 48
47. Leaves without hyaline point or awn; capsules cylindrical to ovoid, erect to pendulous, exserted 50

48. Hyaline points long, often reaching 1 mm or more; peristome with a basal tube and twisted teeth above; on soil or rocks in dry, sunny, calcareous habitats, often on sand near shores of lakes .... [*Syntrichia ruralis*]
48. Hyaline points short, usually <1 mm or if longer, the margins toothed below the point; peristome without basal tube and teeth not twisted, often on rock 49

49. Plants with short, tuftlike branches; capsules exserted ....
.... *Racomitrium,* see **Grimmiaceae** (p. 143)
49. Plants with unbranched stems or with long branches; capsules excerted (*Grimmia*) or immersed ....
.... *Schistidium* (see couplet 29), and see **Grimmiaceae** (p. 143)

50. Plants with a whitish, filamentous, or cobwebby substance on leaves; on soil on steep banks or in rock crevices. A rare calciphile ....
.... [*Saelania glaucescens*]
50. Plants lacking a whitish substance on leaves 51

51. Leaf margins recurved 52

> Note: one *Pohlia* sp., *P. nutans*, has leaves recurved when dry. See couplet 56

51. Leaf margins plane or incurved 55

52. Capsules sulcate or ribbed when dry 53
52. Capsules smooth or indistinctly wrinkled 54

53. Leaves often reddish or purplish, especially at base; pseudopodia lacking; capsules and setae purplish when fresh, turning brown ....
.... *Ceratodon purpureus* (p. 133)
53. Leaves green to yellowish brown; pseudopodia often present with gemmae attached; capsules and setae yellow or brown, sometimes reddish brown ....
.... *Aulacomnium* (see couplet 34)

54. Gametophytes large, often >1 cm high; leaves broad, 0.5 mm or more wide; on rock in or beside streams ....
.... *Racomitrium aciculare* (p. 148), see **Grimmiaceae** (p. 143)
54. Gametophytes small, mostly <1 cm high; leaves narrow, <0.5 mm wide; commonly on calcareous soil over boulders and cliff ledges, sometimes on rotten logs and stumps .... [*Bryoerythrophyllum recurvirostrum*]

55. Stems and costae red; leaves bluish green; on calcareous substrate or in cedar swamps .... [*Mnium stellare*], see **Mniaceae** (p. 155)

55. Stems and costa red or not; leaves not bluish green; not necessarily on calcareous substrate 56

56. Leaves often 5 mm long or more, margins inrolled when dry; calyptrae remaining attached to setae just below capsule; on moist soil or humus, on shaded banks along creeks, or in swamps .... [*Timmia megapolitana*]

56. Leaves <5 mm long, margins plane; calyptrae not remaining attached to setae .... *Pohlia*

Pohlia

A. Plants with gemmae in upper leaf axils .... B
A. Plants without gemmae C

B. Plants with narrow gemmae .... *P. annotina* (p. 111)
B. Plants ovoid gemmae .... *P. bulbifera* (p. 111)

C. Plants yellowish green to dark green; leaf margins recurved; capsules elongate; on soil, humus, rotten logs and stumps in woodland clearings, sometimes in bogs .... *P. nutans* (p. 113)
C. Plants whitish green; leaf margins plane; capsules elongate or short and nearly as broad as long .... D

D. Leaves glossy, often with opalescent patches; capsules elongate; on soil or humus, often on shaded rock ledges or in crevices of cliffs, sometimes on rotten logs .... *P. cruda* (p. 112)
D. Leaves dull, lacking opalescent patches; capsules short, nearly as broad as long; on mostly wet soil in exposed and disturbed habitats ....
.... *P. wahlenbergii* (p. 114)

QUICK LOOK AT FAMILIES BY HABITAT

This section isn't really a key, but more a guide to help narrow the field of possibilities and make some new friends. See also the key to families and the family descriptions following this quick look. Some mosses show a marked loyalty to their primary habitat and substrate, and ecological characteristics can be very helpful in their identification. Then again, some species are generalists and can show up in the strangest places. We'll focus here on the most likely places to find a particular family with no promise of including all possibilities. Please remember to check the quick look at pleurocarpous mosses if your sample isn't clearly acrocarpous. Habitats are listed from wet to dry.

Aquatic

This is largely pleurocarp territory; however, certain *Fissidens* spp. may be found submerged. They have highly distinctive two-part leaves, but the aquatic species appear to be infrequently collected. See the Very Wet habitats for species occasionally submerged.

Bogs

Ombrotrophic bogs, i.e., bogs with minimal groundwater flow and that receive their nutrients from the sky are the province of the peat mosses (Sphagnaceae), but some acrocarpous mosses may be found here.

Splachnaceae: Grows on decaying animal matter or dung. Uncommon.

Polytrichaceae: Grows among *Sphagnum, Polytrichum strictum* is upright, slender, with stiff, folded-over leaves.

Very Wet, Occasionally Submerged, Mucky

Aulacomniaceae: A likely candidate, often easily identified by clusters of gemmae; *Aulacomnium palustre*, ribbed bog moss, is most common and widespread.

Bartramiaceae: *Philonotis fontana* with upright, parallel, red stems and well-spaced leaves is common in damp mucky places. Its asymmetrical, globose capsules are distinctive but often not present.

Mniaceae: *Rhizomnium* and *Mnium* species are particularly likely. Look for large, often bordered leaves and drooping capsules.

Very Wet, Occasionally Submerged, Rock

Grimmiaceae: These are generally dark, blackish-green species often on sporadically inundated rock. If capsules are present, they are upright, either immersed or exserted.

Fissidentaceae: With distinctive flattened, two-part leaves, often found on rocky lakeshores.

Forest Floor

Mosses are more common and conspicuous in conifer forests where fallen leaves don't cover them every year, and where soil pH is often acidic, a condition that makes life more difficult for some of their competitors—the vascular plants. We think of Sphagnaceae (peat moss) as a family of bogs and boggy places, but some species thrive in wet or seepy depressions in forested sites. They are keyed and discussed in the *Sphagnum* section.

Polytrichaceae: Upright, unbranched, often several centimeters tall with divergent leaves when moist, this family has members in many habitats. *Polytrichum* spp. and *Atrichum undulatum* s.l. would be good places to start.

Dicranaceae: Look for the falcate-secund leaves, often with light green tips on darker green plants, growing in neat tufts on soil or rotting wood.

Buxbaumiaceae: The most common member of this family, *Diphyscium foliosum*, is often found on clayey banks along streams, trails, or dirt roads. It is best recognized by its teardrop-shaped capsules on inconspicuous, very short setae, surrounded with fine-tipped perichaetial leaves.

Bryaceae and Mniaceae: Both have drooping capsules, and leaves that often show a border of some sort.

Fissidentaceae: Several species of *Fissidens* with distinctive, flattened, two-part leaves are found on soil.

Bartramiaceae: *Bartramia pomiformis* is often found on soil over rock in ledge sites.

Leucobryaceae: *Leucobryum* spp. grow in neat whitish tufts on the forest floor, often along trails, occasionally on rotting wood or tree bases.

Rotten Stumps and Logs or Humus

As wood decays into humic soil, almost any of the forest floor species can show up, but a few species are rotten-wood specialists.

Tetraphidaceae: This short list is led by *Tetraphis pellucida*, a small species that can be found quite reliably on stumps in advanced decay.

Dicranaceae: Two small *Dicranum* species, *D. flagellare* and *D. montanum,* are also likely prospects. As with the species of waste places, these mosses are on an inherently unstable substrate and constantly need to find new homes. *T. pellucida* and *D. flagellare* produce both capsules and conspicuous vegetative propagules.

Buxbaumiaceae: An unusual and uncommon humic soil species is *Buxbaumia aphylla*, a moss without apparent leaves, and with a large teardrop-shaped capsule.

Leucobryaceae: See Forest Floor.

Tree Bases (up to ≈0.5 m above ground)

Any of the rotten stump and log species, tree trunk species, and some soil species can occasionally be found on tree bases, particularly the lower and more horizontal parts of the tree base; however, as with the aquatics, this is really pleurocarp territory, and acrocarpous species are not common.

Tree Trunks (generally above ≈0.5 m above ground)

Very few species can stand the desiccation encountered on a tree trunk well above the forest floor, but a few exist, and for the most part they're not too difficult to identify, at least to genus.

Orthotrichaceae: A good place to start, this family includes the common neat tuft–forming *Ulota* spp., and the *Orthotrichum* genus of less-neat tree-tuft mosses. All the Orthotrichaceae are good capsule producers and have ribbed capsules with hairy calyptrae.

Dicranaceae: *Dicranum viride*, the only tree-trunk *Dicranum*, favors sugar maple and cedar and is easily identified by substrate and broken leaf tips. Unfortunately it's rather uncommon.

Waste Places

This large category includes road and field edges, gravel parking areas, campsites, gardens, and generally any disturbed, usually sunny site. Waste places are acrocarp territory. Most of these species are very good capsule producers, presumably because the ephemeral nature of their substrate forces them to find new locations frequently.

Bryaceae: With white leaf tips, *Bryum argenteum* is often found in sidewalk cracks. This is not one of the reliable capsule producers—maybe because of the traffic! *Pohlia* spp. are also good waste-place possibilities.

Dicranaceae: *Dicranella heteromalla* has a typical *Dicranum* look (long, narrow falcate-secund leaves) and yellow setae. *Trematodon ambiguus*, with long-necked and gracefully curved capsules, favors relatively dry, disturbed, often roadside soil. Most of the species in this large family are found on forest soils, not in waste places.

Ditrichaceae: *Ceratodon purpureus* creates large purple swards on disturbed soil by the color of its setae and capsules. *Ditrichum pallidum* looks like a small, silky, yellow-green *Dicranum* sp. (it was once in the genus *Dicranum*).

Funariaceae: *Funaria hygrometrica* produces many large, distinctive, asymmetrical, nodding capsules. *Physcomitrium pyriforme*, with an often prolific capsule array, is common in gardens in spring.

Polytrichaceae: *Polytrichum commune* and *P. juniperinum*, both with the look of small unbranched clubmosses, are likely to be encountered on disturbed, more or less sunny sites. *Pogonatum pensilvanicum* grows from greenish-looking soil on roadside or trailside banks with inconspicuous leaves and setae and upright capsules looking at first glance as though they rise from the soil.

Pottiaceae: This family is generally found on calcareous sites. The most common waste-place members of the family might be *Barbula unguiculata*, *Tortula truncata*, and *Weissia controversa*, all of which produce distinctive capsules.

Rock

Most of the acrocarpous species that grow on rock are dark green, even blackish green, and most have immersed capsules, or capsules slightly exserted on a short seta.

Andreaeaceae: The granite mosses are reddish brown to black-brown and grow on exposed noncalcareous rock.

Dicranaceae. A species with falcate-secund leaves might be the rock *Dicranum*, *D. fulvum*.

Grimmiaceae: All three of the genera included from this family are rock dwellers. *Racomitrium* is the odd one here, with longer setae than the rest of the family, and the setae originate from the tip of a branch rather than the tip of the main stem as with the rest of the acrocarps.

Orthotrichaceae: This family of mostly tree trunk species has a couple of possibilities. *Ulota hutchinsiae* is found on both acidic and calcareous rock, and *Orthotrichum anomalum* is largely a species of calcareous rock.

Pottiaceae species that might be found on calcareous rock or soil over rock are often yellow-green and with distinctive capsules.

Dry, Open Soil

Many species of open dry soil are the same species found on waste places. Dicranaceae, Polytrichaceae, and Funariaceae are good bets, with Pottiaceae in calcareous sites.

KEY TO FAMILIES

See family descriptions following for clarification or if a choice is unclear. Adapted from Grout 1924.

1. Plants whitish or gray-green, growing in dense cushions or tufts on soil or rotten wood in shaded forests..Leucobryaceae
1. Plants any color other than above, may have whitish tips, but not glaucous throughout 2

2. Leaves in 2 rows with the edges of the leaves toward the stem, costate or not. 3
2. Leaves spirally inserted around the stem, almost always costate (family Andreaeaceae is the exception) 4

3. Plants of various sizes; leaves costate, folded into 2 laminae, clasping the stem.. Fissidentaceae
3. Plants small, on soil in dark places, with ecostate leaves in 1 plane, divided incompletely; notable for its luminous protonematal mat. ....Schistostegaceae

4. Plants blackish to dark reddish brown, or if green, only so at growing tips; leaves opaque or nearly so; on trees or rock. 5
4. Plants green to light yellow-green, or if blackish, growing on soil.................7

5. Plants small, dark, easily crumbled, primarily on rock in alpine or subalpine habitat, with capsules dehiscing into 4 valves .... Andreaeaceae
5. Capsules dehiscing by a round opening surrounded by 16 teeth, on rock or tree 6

6. Plants on rock, often with hoary appearance from leaf tips lacking chlorophyll; peristome with teeth not reflexed when dry; calyptrae naked..... ....Grimmiaceae
6. Plants mostly on trees (see rock species *Orthotrichum anomalum* and *Ulota hutchinsiae* for exceptions); leaves without whitish tips; peristome teeth usually reflexed when dry; calyptrae hairy with hairs growing upward........... .... Orthotrichaceae

7. Plants on rotting wood, with capsules having 4 teeth, or if no capsules present, with gemma cups..Tetraphidaceae
7. Plants not as above 8

8. Plants with globose capsules and if capsules are long stalked, with no apparent leaves, or if short stalked, immersed in filiform perichaetial leaves .... Buxbaumiaceae
8. Not as above 9

9. Plants with 16 capsule teeth split to base and twisted; leaves usually twisted when dry; often in calcareous habitats. Exceptions: *Tortula truncata* lacks teeth, and *Weissia controversa* has 16 short teeth...............................Pottiaceae
9. Plants with capsules cylindrical or urn-shaped, but not globose, with 16–64 straight (i.e., not twisted) teeth, or teeth lacking; leaves twisted when dry or not; plants of various habitats (for plants with gemmae on stalks go to couplet 17) 10

10. Large upright plants on soil; calyptrae hairy with hairs growing downward (genus *Atrichum* is the hairless exception); peristome with 32–64 teeth, often attached to a drumlike membrane covering the capsule opening; with numerous vertical lamellae on the upper leaf surface .... Polytrichaceae
10. Small to large plants on various substrates; calyptrae without hairs; peristome with 16 teeth or peristome lacking, opening not closed by a membrane attached to teeth; no lamellae on leaf surface 11

11. Capsules urn-shaped and lacking a peristome...
....*Physcomitrium pyriforme* in Funariaceae
....and *Pottia truncata* in Pottiaceae
11. Capsules not urn-shaped and with 16 peristome teeth.................................. 12

12. Capsules with a large swollen hypophysis larger and more conspicuous than the urn; on dung or animal remains, rare .... Splachnaceae
12. Not as above 13

13. Capsules plicate when dry and empty; leaves without a border visible with a hand lens.. 14
13. Capsules smooth or only slightly wrinkled when dry and empty; leaves often bordered with clear or colored border visible with a hand lens 18

14. Plants with globose to subglobose pyriform or urn-shaped capsules 15
14. Plants with cylindrical capsules.. 16

15. Plants with globose capsules.. Bartramiaciae
15. Plants with asymmetrical pyriform capsules or with urn-shaped capsules (see couplet 11) ....Funariaceae

16. Plants with leaves long and narrow, usually >10× as long as wide, often falcate-secund; capsules straight to curved.................................... Dicranaceae
Ditrichum spp. in Ditrichaceae will key here
16. Plants with leaves <10× as long as wide... 17

17. Plants frequently producing gemmae on stalks (*Aulacomnium heterostichum* is the exception)..Aulacomniaceae
17. Plants without gemmae on stalks... Ditrichaceae

18. Capsules pendent and usually with a noticeable neck; leaves frequently more crowded toward the stem tips ....Bryaceae
18. Capsules may be inclined, sometimes pendent, but not with a noticeable neck..Mniaceae

Families Not Included Above (see the family descriptions)

Encalyptaceae species rarely produce sporophytes in our range.
Seligeraceae has only 1 species (*Blindia acuta*) included in this book.

FAMILY DESCRIPTIONS

Andreaeaceae (Granite Mosses)

Genus included *Andreaea*

Substrate Exposed noncalcareous rock, often alpine or subalpine, frequently in a successional situation

Capsules, as shown to the right, are unusual and diagnostic. They are held only slightly above the leaves (shortly exserted), and they open along 4 splits on the sides of the capsules, looking a bit like Chinese lanterns. Teeth: none

Bry. Eur. Developing capsules.

Plants are small, reddish brown to black, growing in brittle tufts, very little branched.

Three species occur in our area. All are discussed and 2 are illustrated. One of the illustrated species has no costa, an unusual characteristic in Acrocarps.

The genus is in order Andreaeales, which contains only 1 family and 1 genus. This and *Tetraphis* are the only genera discussed in the Pleurocarp and Acrocarp sections that are not in class Bryopsida.

Because of their dark color and rock habitat, they are most likely to be confused with the Grimmiaceae species, which are generally larger and with very different capsules.

Aulacomniaceae (Ribbed Bog Mosses)

Genus included *Aulacomnium*

Substrate Soil in moist woods, along drainage ditches, in bogs

Capsules are upright to curved and inclined, cylindrical and with 16 teeth. They are ribbed when dry and empty, thus the common name for the group. Our most common species produce gemmae on stalks, as shown at right.

Left, typical capsule; center, *A. palustre* gemmae stalk; right, *A. androgynum* gemmae stalk

The plants are yellow-green to green, medium to large, stems often tomentose. Leaves are lanceolate to elliptic, unbordered, and with a strong costa ending near the tip.

Five species are possible in our area, but 2 uncommon alpine species are not covered. Of the 3 covered species, 1 produces capsules commonly (*A. heterostichum*), and 2 produce capsules infrequently, reproducing primarily from the gemmae clusters illustrated (*A. palustre* and *A. androgynum*). The gemmiferous species are dull owing to the presence of papillae on their leaf cells, whereas our most likely capsule producer has smooth cells and is shiny.

Bartramiaceae

Genera included *Bartramia, Philonotis*

Substrate *Bartramia pomiformis,* the only species in its genus in our area, prefers soil over rock or in rock crevices in cliff habitats. *Philonotis fontana,* which prefers moist soil on the banks of pools or streams, is by far the most common of the 3 species of that genus that might be encountered.

Left, *B. pomoformis;* right, *P. fontana,* slightly dried

The spherical capsules, both of which have 16 teeth and wrinkle when dry, are a good uniting factor for the family. Unfortunately, *Philonotis* is a less reliable producer of capsules than the fecund *Bartramia*. *Philonotis* has some very nice ID characters, however, so you should familiarize yourself with its species page.

Plagiopus, not illustrated, is another genus in this family that may be found in the northern part of our range on calcareous substrates. It has capsules typical of the family and is distinguished by its triangular stem in cross section.

Bryaceae

Genera included *Bryum, Leptobryum, Pohlia, Rhodobryum, Rosulabryum*

Substrate Soil

Bryaceae capsules, the best family character, have 16 teeth, droop, are not wrinkled when dry, and frequently have a well-defined neck, particularly when dry.

Bry. Eur. *Rosulabryum capillare* capsules.

The leaves, often larger and more crowded toward the stem tip, frequently have a noticeable border, sometimes darker than the rest of the lamina, and occasionally an excurrent costa.

This is a large and complex family with 31 genera and 945 species worldwide. Included are the most common representatives of the 5 genera likely to be encountered in nontropical eastern North America; however, it is important to remember that more than a dozen *Pohlia* spp. and 2 dozen *Bryum* spp. have been recorded as possible (Crum and Anderson 1981).

Of this difficult family with species often lacking sporophytes, Bruce Allen (2005) said, "Sterile material sometimes cannot be determined even to genus without extensive prior experience with the family."

Buxbaumiaceae

Genera included *Buxbaumia, Diphyscium*

Substrate Soil or rotten wood

Capsules large and conspicuous on otherwise small, inconspicuous plants. *Diphyscium foliosum* capsules lack a seta (or appear to) and are surrounded by filiform perichaetial leaves, while *Buxbaumia aphylla* has a long seta and no apparent leaves.

A. J. Grout 1924: "A most peculiar and fantastic family."

Bry. Eur. Left, *D. foliosum*; right, *B. aphylla.*

Some authors separate these two genera into their own families and combine them into subclass Buxbaumiidae.

Two common members of this family occur in our coverage area, and both are illustrated.

Dicranaceae

Genera included *Dicranella, Dicranum. Oncophorus, Paraleucobryum, Trematodon*

Substrate Soil or rock, decaying wood, rarely trees

The capsules are cylindrical, upright and straight, to somewhat curved and inclined, not usually pendent as in the Bryaceae, held well above the leaves on an elongate seta, and either smooth or ribbed when dry. The 16 peristome teeth are split about halfway to the base (except in *Trematodon*, where the 16 teeth are not split), and are often reddish.

Some representatives: left, *D. scoparium*; middle, *D. flagellare*; right *O. wahlenbergii*

Plants are small to large, little branched. Leaves are costate, usually long and narrow with sheathing bases, frequently falcate-secund, often >10× as long as wide, contorting when dry or not.

Four of the five genera included here are treated as "*Dicranum* and Friends." *Oncophorus wahlenbergii* is treated separately.

Left, *O. wahlenbergii*; right, a generic *Dicranum* sp.

Crum and Anderson (1981) said of this family that it is "too large to be characterized in any definitive way." The *Dicranum* genus, the most common of the family, is treated here quite extensively. The other 4 genera treated are much less diverse and are represented by 1 species each. Several less common genera in this family are omitted.

See a more complete discussion on page 119.

Ditrichaceae

Genera included *Ceratodon, Ditrichum*

Substrate Soil, usually disturbed

This family is closely related to the Dicranaceae, and its uniting characteristics are mostly microscopic; for example, the 16 teeth are split, slightly differently from the way they are split in Dicranaceae, and the alar cells, usually differentiated in the Dicranaceae, are undifferentiated here. Oddly, each of the species illustrated is first recognized by the copiously

C. purpureus   *D. pallidum*

produced setae, red in the case of *Ceratodon purpureus*, and straw in *Ditrichum pallidum*.

Capsules are exserted and held well above the leaves.

Two other genera in this family, *Saelania* and *Distichium*, are uncommon but found from Maine north to the Canadian Maritimes and west to southern Ontario. Both genera are included in the Key to Species, but are not illustrated.

Encalyptaceae

Genus included *Encalypta*

Substrate Calcareous soil

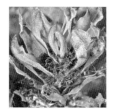

Capsules are distinctive, but the species most likely to be encountered in our region doesn't produce them.

The plants have leaves spatulate to lingulate and very dull and opaque looking, from multiple papillae on the leaf cells. Leaves contort when dry. Dense clusters of brood bodies (illustrated at right) are produced by *Encalypta procera*.

E. procera with brood bodies

The family name is from the capsules' resemblance to a candlesnuffer—see species page for *E. procera*.

This family has only 1 genus and 3 possible species in our area. *E. procera*, the largest and most common in an uncommon group, is illustrated.

Fissidentaceae

Genus included *Fissidens*

Substrate Rock and soil

The best distinguishing characteristic of this family is the unusual leaf construction. Each leaf has 2 connected parts and leaves grow in 2 rows, giving the plants a very flattened appearance.

Capsules are *Dicranum*-like, with 16 teeth split halfway to their base.

This single genus family is unusual in many respects— please see additional discussion starting on page 137, including illustrations of details of genus *Fissidens*.

Bry. Eur. *Fissidens* sp. leaf.

Funariaceae

Genera included *Funaria, Physcomitrium*

Substrate Disturbed soil

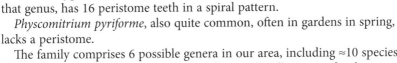

The family comprises primarily annual plants with soft, oblong-lanceolate, or elliptic to obovate, costate, leaves. Copiously produced capsules are urn-shaped, pyriform, or subglobose and asymmetrical.

F. hygrometrica   P. pyriforme

Funaria hygrometrica, our common species in that genus, has 16 peristome teeth in a spiral pattern.

Physcomitrium pyriforme, also quite common, often in gardens in spring, lacks a peristome.

The family comprises 6 possible genera in our area, including ≈10 species; however, the two very common species illustrated will account for the vast majority of collections.

Grimmiaceae (Rock Mosses)

Genera included *Grimmia, Racomitrium, Schistidium*

Substrate Rock

Capsules are cylindrical and symmetrical, not narrowed at the neck, with 16 teeth, and are immersed in leaves (*Schistidium*) to slightly exserted (*Grimmia*), to long stalked (*Racomitrium*). *Racomitrium* produces setae and capsules from the tip of a branch (cladocarpous), and is categorized as pleurocarpous by some authors.

Left, *Schistidium* (Bry. Eur.); middle, *Grimmia* (Ic.); right, *Racomitrium* (Bry. Eur.)

Plants are dark green to blackish, usually in dense cushions. Leaves are lanceolate to ovate-lanceolate with a strong costa frequently ending in a hyaline awn.

See a more complete discussion of this family on page 143.

See also Hedwigiaceae in the Pleurocarp section and its sole representative in eastern North America, *Hedwigia ciliata*. This rock moss is keyed as a pleurocarp though it has characteristics of both acrocarpous and pleurocarpous mosses.

Leucobryaceae

Genus included *Leucobryum*

Substrate Soil, rotting wood, occasionally tree bases

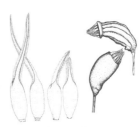

These very common trailside plants grow in distinctive cushions that can be several inches to 1 or 2 feet in diameter with leaves lacking chlorophyll in some cells, giving them a glaucous, or whitish green appearance. The leaves are opaque, fleshy, rolled longitudinally almost into tubes (subtubulose), and costate, though this isn't obvious since the leaves are essentially all costa.

Left, *L. glaucum* leaves; middle, *L. albidum* leaves; right, *L. albidum* capsules (Grout)

Capsules occasionally produced with 16 teeth, divided about halfway to the base. There are only two species in our range and both are discussed.

Mniaceae

Genera included *Mnium, Plagiomnium, Pseudobryum, Rhizomnium*

Substrate Soil

You wouldn't know it from the diversity shown in this leaf illustration (right), but these plants have a "look"

A Mniaceae leaf sampler. Left to right: 1–2, *Plagiomnium*; 3, *Rhizomnium*, 4–5, *Mnium*; 6–7, *Pseudobryum* (Grout 1924).

that you'll get to know. Capsules are much like the Bryaceae, but with less or no neck, and with respect to the Bryaceae, the leaves are generally larger and wider and often show a differentiated border, sometimes in a darker color than the rest of the leaf (see *Rhizomnium*). *Plagiomnium* and *Mnium* leaves are toothed, *Rhizomnium* and *Pseudobryum* leaves are entire. Antheridial cups are common in this family and Polytrichaceae.

Generic Mniaceae capsule (Bry. Eur.)

This family comprises ≈20 species east of the Mississippi. The dichotomous key covers 12 species, and the species pages illustrate what I hope are the most common 7 species, with representatives from all 4 genera.

See a more complete discussion of this family on page 155.

Orthotrichaceae

Genera included *Orthotrichum, Ulota*

Substrate Tree trunks primarily, rock occasionally

Usually small, dark plants with leaves more or less lanceolate and strongly costate. This family has some of the look of the Grimmiaceae, but usually with a tree trunk habit, and leaves more likely to be contorted. As with the Grimmiaceae, the capsules are often immersed to short exserted, but unlike that family, the capsules have hairy calyptrae, are more elongate, and are ribbed when dry. The peristome consists of 16 rather short teeth that are nearly always reflexed when dry and often united in pairs, so it may appear to have 8 teeth.

U. crispa capsules with calyptra and empty

Crum and Anderson (1981) listed 9 genera and a possible 36 species within this complex family. Many of the species are rare and unlikely to be encountered, but some are more widely distributed. The few very common species discussed here will reasonably represent the family.

Polytrichaceae

Genera included *Atrichum, Pogonatum, Polytrichastrum, Polytrichum*

Substrate Soil

Top, *P. commune* leaf showing lamellae and teeth; bottom, *P. juniperinum* showing lamellae covered by the leaf lamina

These are large, upright, generally unbranched plants, with long, divergent (when hydrated) leaves with clasping bases. They often have the look of a small clubmoss or conifer, or in the case of the *Pogonatum* genus, a small *Agave* plant. Their height, which in some species can reach several inches, is facilitated by a vascular bundle in their stems allowing transport of water and nutrients. A defining feature of this family is the presence of rows of tissue (lamellae) on the upper surface of the leaves. These lamellae may cover the entire surface of the leaf, or just a few rows above the costa may occur. In many species, the lamellae render the leaves opaque, fleshy, and stiff, unlike most mosses with their thin, translucent leaves usually one cell thick.

Left to right: *Atrichum* sp. (Bry. Eur.), *Polytrichum commune* (M.V.T.), *Polytrichastrum ohioense* (M.V.T.), *Pogonatum dentatum* (M.V.T.)

Our species are usually dioicous, the male plants producing antheridia in conspicuous flowerlike "splash cups." Calyptrae of most are covered with dense tomentum (*Atrichum* spp. are the exception). Capsules of *Atrichum*, *Pogonatum*, and *Polytrichastrum* have 32 teeth; *Polytrichum* capsules have 64.

Polytrichum splash cup

This is a common and conspicuous family, and all genera in our coverage area and all species likely to be encountered are discussed.

See a more complete discussion of this family on page 170.

Pottiaceae

Genera included *Barbula, Tortella, Tortula, Weissia*

Substrate Primarily disturbed soil, with a marked preference for calcareous substrate

Pottiaceae capsule pairs, left to right: *Barbula, Tortella, Tortula, Weissia*. Drawing on left by the author, all others from Bry. Eur.

Small to medium-sized plants, unbranched, and usually growing in dense tufts. Leaves are long, narrow, and with well-developed costae. Most species have leaves contorted when dry. Leaf cells are usually multipapillose (*Tortula* is the exception, with smooth cells), giving a dull appearance. Capsules are lacking teeth or have 16 teeth, frequently split to the base and twisted. With the distinctive if diverse capsules, we are fortunate that most species in this family are reliable capsule producers.

This is a huge and diverse family with more than 100 genera worldwide, and 30 in North America east of the Mississippi River. The distinct preference for calcareous substrate makes them less than common in much of our coverage area.

Schistostegaceae

Genus included *Schistostega*

Substrate Gravelly soil in dark places
Very strange small plants that appear to glow in the dark as a result of a reflective protonematal mat. Leaves are flat, ecostate, and two-ranked. Capsules are small, not often produced, rising from perichaetial leaves that are tufted and quite unlike the distichous sterile leaves.

S. pennata
(Bry. Eur.)

This family has 1 genus and 1 species in our area.

Seligeriaceae

Genus included *Blindia*

Substrate Damp rock, frequently in a spray zone
Capsules have 16 undivided teeth.
Small plants with narrow *Dicranum*-like leaves and bulbous, 16-tooth capsules.

This is a small family with only 3 genera in eastern North America. The 2 genera not covered here are reported to be uncommon. This one may be uncommon also—see species page.

B. acuta (Bry. Eur.)

Splachnaceae

Genus included *Splachnum*

Substrate Dung or animal remains, often in bogs
These are unusual mosses on an unusual substrate and with unusual capsules. The plants are short stemmed and reliably produce capsules that are upright and with an enlarged (greatly so in some species) hypophysis (lower section where the capsule meets the seta), and 16 teeth, reflexed when dry.

S. ampullaceum

This is a family of infrequently encountered mosses, and I represent it in this book with one genus. Two other genera in the family are remote possibilities for collection in eastern North America, *Tetraplodon* and *Tayloria*. None of the Splachnaceae is likely to be encountered south of northern New England, and *Tayloria* is primarily an Arctic genus.

Tetraphidaceae

Genus included *Tetraphis*

T. pellucida (Bry. Eur.)

Substrate Decaying wood (rotten stumps)

Our common species, *T. pellucida*, is distinctive for its 4 tooth capsules (our only family with such), its propensity to produce gemma cups, and its rotten stump substrate. Leaves are costate and quite translucent (pellucid).

Tetrodontium is the only other genus in this family in eastern North America, and its plants are very small, rarely produce capsules, and are not likely to be encountered.

Acrocarpous Species Accounts

Presented alphabetically by family

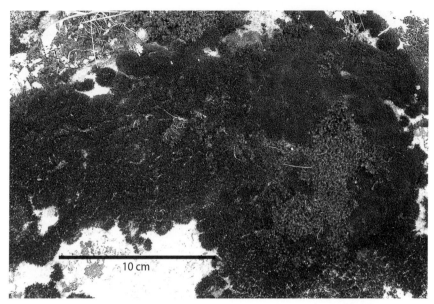

Andreaea rupestris on Cadillac Mountain, Mount Desert Island, ME; dry on the left and moist on the right

Description Our two species of *Andreaea* share a black to red-brown color, small-tuft growth form, tiny leaves, and exposed noncalcareous rock habitat. Like many mosses, they dry out regularly and are able to revive with no ill effects.

Andreaea capsules are very distinct, unique to the genus, and though small, can often be found slightly raised from the dark tufts (next page).

A. rothii, less common than the next, has a strong costa occupying more than half the leaf width at midleaf and elongate, almost falcate-secund leaves.

A. rupestris, the more common species, can be separated from *A. rothii* by its broad, appressed leaves lacking a costa.

Similar Species The two species of *Andreaea* are difficult to separate at first glance. The detail photos to the right show that the leaves look very different. Careful examination with a hand lens should easily separate the two species. If you can find capsules (see next page), they are unique to this genus.

The *Racomitrium, Grimmia,* and *Schistidium* genera are common rock mosses that can be quite dark, but they are green to black-green, lacking the reddish to black aspect of the *Andreaea* spp., and they frequently have hyaline leaf tips or awns, which these granite mosses never have.

Andreaeaceae   Continued on next page

A typical tuft of *Andreaea*, probably *rothii*. Cadillac Mountain, Mount Desert Island, ME.

Range and Habitat Throughout our range on exposed noncalcareous rock.

Name The genus name is for J. G. R. Andreae (1724–1793), an apothecary in Hanover, Germany, who studied the geology and natural history of the Swiss Alps. The species name, *rothii*, is for A. W. Roth (1737–1834), a German botanist who collected this species on boulders on ancient burial grounds. *rupestris* (L) means rock-dwelling, and *crassinervia* (see below), from *crassus* (L) = thick and *nervus* (L) = nerve, refers to the wide costa. Common name: granite moss.

Note [*A. crassinervia*], not illustrated here, is rare or possibly underreported in our region. It is so similar to *A. rothii* that separation in the field is not possible. Its costa is broader than in *A. rothii*, occupying the entire leaf width at midleaf, and is excurrent versus the usually percurrent costa of *A. rothii*.

Andreaea capsule, 24×

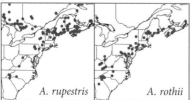

Bry. Eur.

Forest floor habitat with inset of gemmae, 10×. Eagle Hill campus, Steuben, ME.

Description This is a greener, smaller version of *A. palustre* with globose gemma clusters at the tips of bare stalks lacking the tiny leaves present on *A. palustre* gemma stalks. Stem leaves 2–3.5 mm long have a dull look due to microscopic papillae; they have recurved margins, are divergent and usually incurved when moist, lanceolate, with small teeth near the tip, and costate with a strong costa ending just a few cells before the leaf tips. Dioicous and an infrequent capsule producer. If produced, the capsules would be similar to those illustrated for *A. palustre*, another meager capsule producer.

Hand lens view, 4×

Similar Species See *A. palustre*.

Range and Habitat Distribution in the east is largely confined to coastal Maine, the Canadian Maritimes, and around the Great Lakes. Much more common on the west coast of North America. On soil in moist woods or bog margins.

Name *Aulacomnium* is from *aulax* (G) = furrowed, and *mnium*, an ancient Greek name for a moss, referring to a *Mnium*-like moss with furrowed capsules. The species name *androgynum*, from the same root as the English word androgynous, refers to the presence of male and female organs on the same individual. This moss is actually dioicous; that is, its sex organs are on different plants, but early botanists incorrectly thought the gemma clusters contained the antheridia.

A. androgynum

Aulacomniaceae

25 mm

Above, on soil over limestone, Ravena, NY;
right, herbarium specimen

Description Growing in cushions with stems to 4 cm long, either branched or simple, with somewhat complanate, oblong leaves 2–2.5 mm long, strong costa, and obvious teeth in the upper half. The stems and leaves all tend to bend in the same direction.

Stems have dense red-brown rhizoids (tomentum) on lower portions and are ridged because of the decurrent leaf bases. Capsules are ribbed and contracted at the mouth when dry. These plants are monoicous and produce capsules frequently. Without capsules, this species is quite *Mnium*-like. Short stalks with gemmae are infrequently produced, but these are what tie this species to the genus. This *Aulacomnium* is likely to produce capsules, whereas the other two species largely eschew sexual reproduction in favor of producing gemmae.

Similar Species The *Mnium*-like leaves, unbordered, very complanate, strongly costate and obviously toothed, combined with the ribbed capsules should be diagnostic.

Range and Habitat Rare or at least uncommon in Maine, but more common in the southern portion of our range. Habitat descriptions vary: Allen (2005)

A. heterostichum

said, "on soil of stream and trail banks in shaded places"; Crum and Anderson (1981) said, "in rather dry, open deciduous woods"; and Grout (1903) said, "common on rich moist soil (not wet) in woods, especially about the bases of trees."

Name *hetero* (G) = other or different, and *stichus* (G) = a row or line, referring to the complanate leaf arrangement.

Aulacomniaceae

Mount Desert Island, Bar Harbor, ME

Description A fairly large bog moss with stems 4–9 cm long, and with a distinctive bright yellow-green color. Stems show little branching and are covered in a dense reddish tomentum. Leaves, 2–4 mm long, have a dull look due to microscopic papillae, are narrowly lanceolate with small teeth near the tips, have a strong costa ending before the leaf tip, and have recurved margins. They are divergent and slightly incurved when moist, and contorted when dry. The ribbed (when dry) capsules give this species its common name. With such distinctive capsules, it's unfortunate that this species and *A. androgynum* are not frequent capsule producers, reproducing primarily by gemmae. The stalked, terminal, gemma clusters are distinctive. The tiny leaves on stalks leading to gemmae decrease in size until little more than scales.

Fresh   1 mm

Dry

Similar Species The stalked gemmae of *A. androgynum* begin abruptly as a ball on naked stalks versus this species, which has gemma stalks with leaves decreasing in size toward the tip, a cluster of tightly packed gemmiferous leaves.

A. palustre

Range and Habitat Common in bogs or in seepy rock crevices.

Name *palustre* (L) = marshy, referring to the habitat. Common name: ribbed bog-moss.

Aulacomniaceae

20 mm

Arrowsic, ME

Description With such distinctive capsules, the long costate leaves don't get much attention; when hydrated, they have a "one way" divergence shown in the photo above that is hard to describe, but usually easy to recognize; when dry, they are wildly but not tightly contorted. The narrow leaves are finely toothed in the upper 2/3, and their cells papillate, giving them an opaque look. The reddish stems are covered with brown rhizoids. The capsules darken and the peristomes become red as they age. The photographs above and to the right were taken in spring showing new capsules, and the photograph on the lower right shows the early summer look.

New capsules, 5✕

Similar Species See *Philonotis fontana*, another member of the Bartramiaceae family with distinctive capsules. *Dicranum* gametophytes are similar but have very different capsules.

B. pomiformis

Range and Habitat Throughout our range on soil over rocky ledge on partly shaded cliffs or on trailside or roadside banks.

Mature capsules, 5✕

Name Named by Johannes Hedwig for John Bartram, a noted eighteenth-century Philadelphia botanist. *pomiformis* is from *pomum* (L) = apple. Common name: apple moss.

Photos are from a wet grassy roadside, Hancock County, ME; lower left, typical stems and the occasionally produced antheridial heads that lead to the unusual branching shown lower right

Description A medium-sized bright yellow-green moss usually in dense tufts with stems 2–8 cm high, red, and occasionally showing between well-spaced leaves, and with much red-brown tomentum on lower parts. Note the whorled branching pattern sometimes found in this species but not commonly found in other acrocarpous species. These branches radiate from last year's antheridial head. Leaves are 1–2 mm long, costate, acuminate, long-pointed, erect-spreading when wet to not much changed, and with tips slightly twisted when dry, appearing dull because of the rough surface caused by raised cell ends (prorulae). Its unusual capsules are globose, asymmetrical, and ribbed when dry. Dioicous, producing capsules occasionally.

Similar Species Without capsules this can be confused with *Pohlia wahlenbergii*, which shares its light color, its habitat preference, and its red stem. However, *P. wahlenbergii* leaves are more crowded and appressed when dry and are not long-acuminate, its lower stem lacks the dense red-brown tomentum and doesn't branch in a whorled pattern, and because its leaves aren't prorulate, they are more glossy. The long-pointed leaves of *P. fontana* all tend to point in

the same direction, giving a bit of a windblown look not found in *Pohlia*. See also *Pohlia cruda* and *P. nutans*.

P. fontana

Range and Habitat Common throughout our range on damp streambanks and in wet seepy places.

Name *philos* (G) = loving, and *noter* (G) = wet, for the habitat; *font* (L) = fountain for the wet habitat in case you missed it the first time.

Bartramiaceae

Gravel driveway edge, Nelson, NH, 2×

Description Small, in dense tufts with a frosted look caused by white leaf tips. When found in its characteristic sidewalk crack environment, the silver may be less noticeable because of trampling and urban grime. The tiny (<1 mm long) costate leaves have white hair tips and are tightly appressed, giving a julaceous or ropy appearance. Usually sterile, but when this moss is being overtopped by another species it has unrestrained sex and produces many nodding capsules, as shown in the bottom photo. Capsules begin green and become red. Setae are red.

Branch tips, 5×

Similar Species Generally a fairly straightforward ID.

Range and Habitat Very common and found throughout the world. Other than in sidewalk cracks, this pollution-tolerant moss is happy in any sunny, sterile, and gravelly or sandy spot where there isn't much competition. Golf course greenkeepers are not fans of this tough little moss.

Capsule cluster

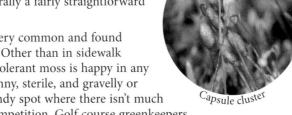

B. argenteum

Name *Bryum* is from *bryon*, an ancient Greek word for moss or lichen, and *argenteum* (L) = silver. Common name: silver *bryum*.

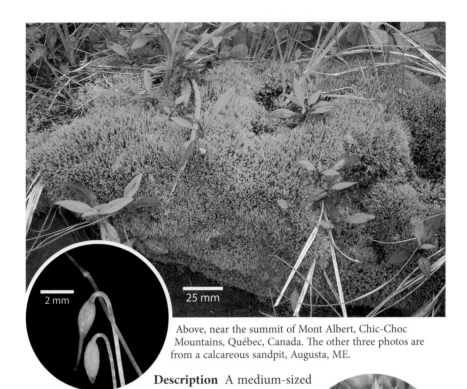

Above, near the summit of Mont Albert, Chic-Choc Mountains, Québec, Canada. The other three photos are from a calcareous sandpit, Augusta, ME.

Description A medium-sized bright to dark green to reddish-brown moss with red upright stems 1–5 cm long covered at least on lower parts with brown tomentum. Lanceolate leaves 2.5–4 mm long have a prominent reddish costa, reddish (less reliably so than the costa) recurved margins, are wide spreading when moist, contorted when dry, extending down along the stem (decurrent), and are widely spaced on the stems to more crowded as in the photos on the right.

Similar Species The reddish costa and leaf margins, divergent and frequently widely spaced leaves, and habitat seem to set this apart.

Range and Habitat Fairly common throughout our range, on wet soil, or on seepy rock outcrops, frequently in open areas.

B. pseudo-triquetrum

Name *Pseudotriquetrum* is less than clear. Crum and Anderson (1981) suggested that it may refer incorrectly to 3-ordered or ranked leaves, but the true meaning remains a mystery.

Bryaceae

On shaded rock near foundation of Squirrel Point Lighthouse, Arrowsic, ME; below, capsule, on soil in a rock crevice at Marshall Point Lighthouse, Port Clyde, ME

Description A small, yellow-green moss usually in dense, frequently extensive tufts with unbranched stems <1 cm tall; leaves long (4–5 mm), hairlike, erect-spreading when wet to wavy when dry, costate with costa occupying most of the leaf width; dioicous and usually producing many 16-tooth, pendent, occasionally horizontal, pyriform capsules on flexuose yellow to reddish setae.

Similar Species The long-necked capsules and hairlike leaves are distinct.

Range and Habitat Found throughout North America on disturbed soil, burned areas, greenhouses, and in coastal locations on soil over rock.

Name *leptos* (G) = slender, plus *bryum* for a thin-leaved *Bryum* relative. *pyr* (L) = a pear, for the pear-shaped capsules.

L. pyriforme

Bryaceae

Above and lower right, *Pohlia annotina*, Arrowsic ME, on the margin of a gravel driveway; upper right inset, *P. bulbifera*, Portland, ME

Description Both species are small, <2 cm tall, with unbranched reddish stems; leaves 1–2 mm long, lanceolate, costate, flat, and held loosely away from the stem. Both have yellowish gemmae in the upper leaf axils. *P. annotina*, the less shiny plants, have elongated gemmae, while *P. bulbifera* is more glossy, and with ovoid, football-shaped gemmae. Both species are dioicous, producing horizontal to pendulous capsules with distinct necks.

Similar Species [*P. proligera*], less common than these two species in our range, is very similar to *P. annotina*, with gemmae more filamentous. It may require microscopic inspection to separate the two.

Range and Habitat *P. bulbifera* is found in the northern portion of our range south only to northern New England, whereas *P annotina* is found throughout our range, south to Tennessee and North Carolina. On damp disturbed soil, frequently at the margins of dirt roads, trails, and ditches.

Name *Pohlia* is for Johann Pohl (1746–1800), doctor and professor in Dresden. *annotinus* (L) = of the previous year, meaning annual. Since these plants aren't annuals, nor do they produce recognizable yearly growth markers, the reason for use of the word is uncertain. *bulbus* (L) = bulb-shaped, for the gemmae.

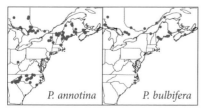

P. annotina   *P. bulbifera*

Bryaceae

Images left and bottom right are from herbarium collections; above right, Hermann Schachner, courtesy of Wikimedia Commons.

Description Light green to whitish-green plants in loose or extensive tufts, with bright red stems to 5 cm long, reddish-brown rhizoids, leaves 2–3 mm long, shiny, costate, sometimes crowded at the stem tip, with margins revolute or plane, and a red costa. Monoicous (occasionally dioicous) and a reliable capsule producer. Capsules 2.5–4 mm long with a neck less than half the capsule length, and yellow to reddish-brown teeth. Operculum (not shown) is bluntly short-conical and lighter in color than the capsule.

The leaves have a waxy surface and are reported to have a lustrous metallic sheen with opalescent patches, in theory making this easy to recognize.

Similar Species Reported to look much like *P. wahlenbergii*, though that species is more likely to be in a very wet place. The capsules are quite different, but *P. wahlenbergii* rarely produces them. This species also resembles *P. nutans*, but that species lacks the distinctive lustrous sheen.

Philonotis fontana is found in wetter locations and has very different capsules.

Range and Habitat Throughout our range on sandy soil, or humus on shaded rocky ledges, or cliffs.

P. cruda

Name *crudus* (L) = raw or fresh, for the lustrous sheen.

On disturbed soil in partial sun, Arrowsic, ME; middle right, dry capsules; bottom right, fresh capsules

Description This common, bright green species has red stems to 3 cm high; leaves lanceolate, 2–2.5 mm long, with a sometimes red costa. The leaves shown above are almost completely the very long filiform perichaetal leaves to 3.5 mm long. Capsules, orange, 2.5–4 mm long, have a well-developed neck.

The fresh capsule, lower right, shows the developing neck.

Similar Species *P. cruda* has a metallic sheen and *P. wahlenbergii* is very white. *P. elongata* is a rare high-elevation species with capsules 3–7 mm long and very long-necked. See also *Philonotis fontana*, which has a similar look but with very different capsule shape, and an unusual branching pattern.

P. nutans

Range and Habitat *P. nutans* is a common species found throughout our range on rotting wood, humic soil, cliff crevices in dense forests, and open places.

Name *nutans* (L) = nodding, for the capsule stance.

Bryaceae

Winslow Ledges, a calcareous outcrop beside the Kennebec River, Winslow, ME

Description Forms dense to sparse, whitish-green tufts 1–3 cm high, red stems occasionally forked, leaves erect-spreading when wet, appressed when dry, 1–2 mm long, lanceolate, costate, with plane margins, and with minor serrations at the tip and entire below. Dioicous, with capsules reported to be ovoid, 1–2 mm long, though rarely produced.

Similar Species *Philonotis fontana* grows in similar habitats, has light-colored leaves and a red stem; however, *P. wahlenbergii* leaves are more appressed when dry and are not long-acuminate, its stem lacks the dense red-brown tomentum and is not branched in a whorled pattern (though it forks), and because the leaves aren't prorulate, they are more glossy. Also, as compared with *P. fontana*, this species doesn't form antheridial heads, and the bulbiform branch tip is unlike *P. fontana*.

Range and Habitat Throughout the Northeast, south to North Carolina, in sunny to partly shaded, seepy habitats.

Name *wahlenbergii* is for Goran Wahlenberg (1780–1851), the Swedish doctor and botanist who collected the type specimen in Norway.

P. wahlenbergii

Bryaceae

Above, on a Native American shell midden, Damariscotta, ME, where the oyster shells provide the calcareous chemistry this species requires; right and below, in rich woods, Mt. Cutler, Hiram, ME

Description Mosses don't get much more charismatic than this. The former and more appropriate name for the plants in our range was *R. roseum*. This is a large moss, with unbranched secondary stems to 3 cm high rising from subterranean horizontal main stems. Leaves 4–9 mm long, costate, and with cuspidate tips are clustered at the stem tip in a floral pattern. Dioicous, with red setae to 3.5 cm long and frequently clustered with as many as 5 setae per stem; capsules pendulous, 4–6 mm.

Similar Species *Climacium dendroides* is also large, has leafless stems and a dendroid look, but it's a much-branched pleurocarp with leaves not clustered at the stem tips.

Range and Habitat Throughout our range on soil or soil over rock in moist, shaded, rich, often calcareous woods.

Name *rhodo* (G) = rose, and *bryon*

R. ontariense

(G) = moss; *ontariense* for an 1897 collection site, probably of the type specimen.

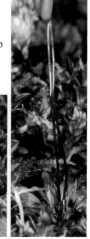

Bryaceae   Ca in part

Winslow Ledges, along the Kennebec River in full sun, Winslow, ME. Note glasses for scale.

Description Crowded, small, dark green to brownish tufts with reddish stems to 1 cm high, stems covered with reddish brown rhizoids (actually filiform vegetative propagules produced in the leaf axils); leaves 1–2 mm long with a strong, frequently reddish costa, slightly twisted at tips when wet and more generally twisted and with recurved margins when dry, hydrated leaves upright to spreading, wide spreading at the tips, giving a flowerlike look from above, dry leaves twist spirally around stem; dioicous, with capsules horizontal to pendent, with 16 teeth and a neck occupying roughly 1/3 the capsule length.

Note thumbnail for scale.

Well-hydrated floret

Similar Species Some species in the *Mnium* genus are similar, but with capsules shorter-necked and not usually pendent, though sometimes horizontal. This is a widespread species with much regional variation.

R. capillare

Range and Habitat Scattered throughout North America on soil in rich woods, soil over rock, or like the sample photographed above, soil over limestone.

Name *capillaris* (L) = pertaining to the hair, for the hairy-looking propagules produced along the stems.

On bare soil along the Squirrel Point Lighthouse Trail, Arrowsic, ME

Description An annual, this species produces its capsules in late fall; they mature in spring and probably begin to disappear by midsummer. These distinctive capsules are 4–5 mm long, reddish brown and on a 5–10 mm seta. Leaves are essentially microscopic.

Range and Habitat Uncommon, but widespread in eastern North America. Crum and Anderson (1981) indicated that it is common in the New Jersey Pine Barrens. A pioneer on depauperate gravelly soils along roads and trails, and on old logs and stumps in open or shaded woods, it is reported to favor burned places.

Similar Species [*B. minakatae*], a rare species more likely to grow on wood rather than soil, is smaller with green or yellow capsules that are cylindrical versus the flattened capsules of *B. aphylla*. *Diphyscium foliosum* has similar capsules, but almost no seta, and distinctive leaves.

Name *Buxbaumia* is for Johannes Buxbaum, the German botanist who described this species in 1712. *A* (G) = without, combined with *phyll* (G) = leaf, indicates that leaves are lacking. Common name: bug-on-a-stick.

B. aphylla

D. foliosum

Buxbaumiaceae

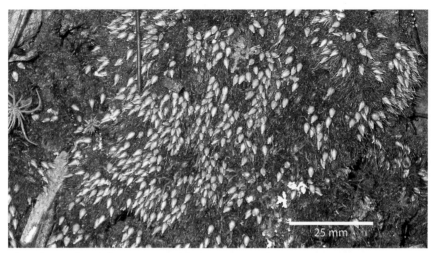

Along the North Traveler Mountain Trail, Baxter State Park, ME

Description A very common and easy to find moss of trailsides and road banks in shaded forests. The plants are dark green with essentially nonexistent stems (1–2 mm long), and leaves ≈2 mm long, lingulate with rounded tip and costa stopping well short of the tip. The perichaetial leaves that surround the capsules end in a long spinelike awn (see right).

The capsules, 2–3 mm long, are composed of 2 sacks, an inner one producing spores, and an outer one providing a bellows effect to eject the spores when struck by a drop of rain or stomped by an insect. The setae are greatly reduced, and capsules appear sessile.

Similar Species *Buxbaumia aphylla*, the only moss with somewhat similar capsules, lacks leaves, has a clearly visible stalk, has larger capsules with a different shape, and is much less frequently encountered.

Range and Habitat Common from southern Québec and southern Ontario to Georgia and Louisiana on shaded banks of damp, bare soil. See map on previous page.

Name *di* (G) = two, and *physi* (G) = bladder, for the double bellows of the capsule. *foli* (L) = leaf and *-osum* (L) = full of, for the leafy perichaetial leaves surrounding the capsules. The specific name serves to distinguish this species from *Buxbaumia aphylla*, with which it once shared a genus. Common name: grain-of-wheat moss.

Note Because of its lack of branching and its capsules produced at the tips of the short stems, I have followed Ireland's (1982) lead and listed this (and *Buxbaumia aphylla*) as an acrocarpous moss, though some authors include it with the pleurocarps.

Buxbaumiaceae

Dicranaceae

GENUS *DICRANUM* AND FRIENDS
THE BROOM MOSSES

The genus *Dicranum* includes several common acrocarpous mosses characterized by long, narrow leaves, usually 10–20× as long as wide, a frequently terrestrial habitat preference, and a strong costa occupying <1/3 of the leaf at the base. The leaves are usually curved (falcate, or sickle-shaped), and they frequently are aligned with their neighboring leaves (secund). This characteristic gives them a combed look, leading to their common name of broom mosses. The spreadsheet (p. 121) includes a few mosses that fit much of the description but are not actually members of the genus. Hence "*Dicranum* and friends."

They typically grow upright, show little or no branching, and have stems covered in a more or less dense mat of rhizoids.

Broom mosses are frequently a large part of the bryological forest floor scene, so it's a good idea to get comfortable with identifying this common cast of characters. Fortunately, the common species are fairly easy to identify.

The capsules shown below are all from *D. scoparium*, beginning at the left with developing "spears" and moving to more mature capsules (top right), then to dry capsules from last season mixed with a few new swords (bottom right).

Dicranum Key

How we got here: These plants have upright, tomentose, unbranched stems of varying lengths; long, narrow, frequently subulate leaves that often are curved and aligned with each other in a nice demonstration of falcate-secund; cylindrical capsules, straight or curved.

Key characteristics to note: Substrate, leaf length, crisped or not when dry, undulate or not, and capsule shape, orientation, and number of setae per stem tip.

1. With straight and erect capsules..2
1. With curved and inclined capsules..5

Straight Capsules

2. With straight flagelliform branchlets interspersed between falcate-secund leaves...*D. flagellare* (p. 123)
2. Without flagelliform branchlets..3

3. Leaves not crisped when dry, many tips broken; on tree trunks............................
....*D. viride* (p. 129)
3. Leaves crisped when dry, leaf tips not broken; on soil or rock..........................4

4. Leaves 5–7 mm long, on rock or thin soil over rock............*D. fulvum* (p. 124)
4. Leaves 2–4 mm long, on humic soil or tree base............*D. montanum* (p. 126)

Curved Capsules

5. Sporophytes single per stem tip..6
5. Sporophytes multiple per stem tip..8

6. Leaves 8–12 mm long, crisped or undulate and twisted when dry..................7
6. Leaves 5–9 mm long, little changed when dry.................*D. scoparium* (p. 128)

7. Leaves crisped when dry, not in a bog...[*D. fuscescens*]
7. Leaves keeled, undulate, and with twisted tips, in a *Sphagnum* bog (see image below)...[*D. undulatum*]

8. Leaves 6–9 mm long, crisped when dry...*D. ontariense*
8. Leaves not much changed when dry..9

9. Leaves 8–12 mm long, not undulate; on soil, usually in a damp conifer forest.
....*D. majus* (p. 125)
9. Leaves 7–10 mm long, undulate; on soil in mixed mesic forest............................*D. polysetum* (p. 125)

See also *Dicranum* and Friends on page 121.

Until recently, *Dicranum* plants with straight capsules were placed in the genus *Orthodicranum* and will be found under that name in most references. *Ortho* is Greek for straight.

D. undulatum collected by Nancy Slack in a *Sphagnum* bog, Mount Desert Island, ME

Dicranum and Friends

Species	Habitat	Leaves, moist	Leaves, dry	Stem rhizoids	Capsules/Comments
Dicranella heteromalla	Soil, sandy, shaded, and disturbed	Falcate-secund, 3–4 mm long	Wavy, but not contorted	None	Oblique mouth capsules, small and silky, yellow seta
Dicranum flagellare	Rotting wood	Falcate-secund to erect-spreading, 2–4 mm	Not much changed	Dense, reddish brown	Straight capsules, distinctive flagellae
Dicranum fulvum	Acid rock, sometimes soil over rock	Falcate-secund to erect-spreading, 5–7 mm	Crisped	Sparse, reddish brown	Straight capsules, very dark lower parts
Dicranum fuscescens	Soil and humus, mainly in coastal forests	Falcate-secund, 8–12 mm	Crisped	Dense, white to reddish brown	Arcuate, inclined capsules, single setae
Dicranum majus	Soil and humus, mainly in coastal forests	Falcate-secund, 8–12 mm	Not much changed	White to reddish brown	Arcuate, inclined capsules, multiple yellow setae
Dicranum montanum	Humic soil and tree base	Erect-spreading to falcate-secund, 2–4 mm	Strongly crisped	Dense, reddish brown	Straight capsules
Dicranum ontariense	Humic soil in coniferous forests	Erect-spreading to somewhat falcate-secund, 6–9 mm	Strongly crisped	Dense, reddish brown	Arcuate, inclined capsules, multiple setae
Dicranum polysetum	Soil, mostly in coniferous forests	Falcate, but not neatly secund, 7–10 mm	Not much changed	Dense, white to reddish brown	Arcuate, inclined capsules, multiple setae, yellow to red, undulate leaves
Dicranum scoparium	Soil, mostly in deciduous forests	Falcate-secund, 5–9 mm	Not much changed	Dense, white to reddish brown	Arcuate, inclined capsules, neat, tight clumps
Dicranum spurium	Soil, forest or open, coastal?	Erect-spreading to incurved, 5–7 mm	Crisped	Dense, reddish brown	Arcuate, inclined capsules
Dicranum undulatum	Hummocks in *Sphagnum* bogs	Erect-appressed to slightly falcate, 5–8 mm	Twisted tips	Dense, reddish brown	Arcuate, inclined capsules, single setae, undulate leaves
Dicranum viride	Tree trunk	Erect-spreading to falcate-secund, 4–6 mm	Erect	Sparse, white to reddish brown	Straight capsules, broken leaf tips common
Ditrichum pallidum (Ditrichaceae)*	Soil, trailsides and roadsides	Falcate-secund, 3–5 mm	Not much changed	None	Curved capsules, small and silky with yellow setae
Oncophorus wahlenbergii	Rotting wood	Long-linear, flexuose, with whitish bases, 4–6 mm	Crisped	White to reddish	Arcuate, inclined, strumose capsules
Paraleucobryum longifolium	Rock, soil over rock. Often high elevation	Falcate-secund, 4–7.5 mm	Not much changed	Reddish	Straight capsules, silky, gray, glossy sheen

Note: Yellow shading indicates a species page.

*Ditrichum pallidum, in a different family, is included here because of its falcate-secund leaves.

Above and right, Black Mountain Trail, Hancock County, ME; below, capsules from an herbarium collection

Description Usually in small, dense tufts of unbranched or once forked stems 1–5 cm long with long-pointed, subulate, falcate-secund leaves 3–4 mm long, giving the appearance of a small silky *Dicranum* sp. The seta is yellow. The dry capsule is deeply furrowed, twisted, constricted, and usually with a bend just below the mouth. In a fresh capsule, the nose is about as long as the capsule itself. Dioicous, and a very reliable capsule producer.

2 mm

Similar Species *Dicranum flagellare* is roughly the same size, but it produces flagellae, grows on rotting wood or very humic soil, and has straight capsules, versus curved (or with a bend) for this species. *Ditrichum pallidum*, another good capsule producer, also has the small silky look with long-pointed *Dicranum*-like leaves and yellow setae, but its dry capsules are small-mouthed without the constriction and bend. See *Dicranum* and Friends (p. 121).

Range and Habitat Throughout North America on shaded, sandy, disturbed places. Very common on trailsides and in high-elevation sites.

Name *Dicranum* refers to that genus, and *-ella* (L) is a diminutive suffix. *hetero* (G) = other, or different, and *omal* (G) = even, for leaves that extend evenly from the different sides of the stem, that is, bend in all directions. Clearly not a very good name for this species with its falcate-secund leaves.

D. heteromalla

Dicranaceae

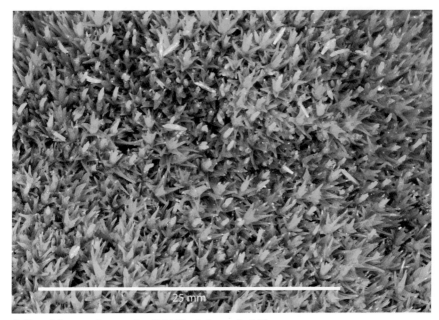

Collected from rotting wood at Bald Head, Arrowsic, ME

Description One of the smaller *Dicranum* species, *D. flagellare* can grow in extensive velvety carpets or in small, tight clumps. The short unbranched stems are tomentose in lower sections with small leaves, 2–4 mm long, falcate-secund to erect-spreading, subtubulose, not much changed when dry, with costa covering <1/4 of the leaf at the base. Usually the leaves are interspersed with straight branches, as shown above and right. These straight, stiff, "flagellate" branches, covered with tiny imbricate leaves, break off easily and can start a new plant. Capsules are straight, upright to inclined.

Similar Species See *Dicranum* and Friends chart, particularly *Dicranum montanum*, which is much more crisped when dry.

D. flagellare

Range and Habitat Found throughout North America from the Arctic treeline south to Central America on very rotten wood or humus in shady locations.

Name *dicran* (G) = pitchfork, for the forked peristome teeth in this genus. *flagellum* (L) = a whip, for the flagellate branches.

Dicranaceae

Chicken rock, along the Squirrel Point Lighthouse Trail, Arrowsic, ME

Description Short unbranched stems form extensive mats on rock or soil over rock in shaded forest settings. Leaves are long and narrow, 5–7 mm long, falcate-secund, and crisped when dry. Setae are yellow to brown; capsules are straight and the operculum has a long snout (below right). Lower parts of the plant look quite dark.

2 mm

Similar Species See the *Dicranum* and Friends chart. The closest look-alike is *Paraleucobryum longifolium*, previously known as *Dicranum longifolium*. *P. longifolium* also grows on rock, and its leaves are about the same length (4–7.5 mm), and are falcate-secund. However, *P. longifolium* is a high-elevation species, it has a gray sheen when dry, its leaves are almost entirely covered by costa versus <1/4 at the base for *D. fulvum*, are little changed when dry versus crisped for *D. fulvum*, and its leaves may have longitudinal striations under hand-lens magnification caused by strips of green cells contrasting with strips of clear cells.

D. fulvum

Range and Habitat Throughout our range, south to Georgia on rock, soil over rock, or occasionally just on soil, or tree trunks.

Name *fulvous* (L) = the tawny color of lions, possibly for the yellowish tint to the tops of the plants.

Dicranaceae

Above left, *Dicranum majus* (1×); center and right, *D. polysetum*, 1× and 3× (note the undulations); below, the larger *D. polysetum* mixed with the smaller and more organized *D. scoparium*, lower right

Description These are big *Dicranum* spp. with dense white to reddish-brown tomentum on stems to 10 cm tall, though usually less. Leaves are 7–12 mm long, undulate (not for *D. majus*), glossy, and falcate-secund, at least at the stem tips. The plant tips tend to point in different directions, resulting in a bit of a disorganized look. They both produce multiple setae from a single stem tip. Setae in *D. polysetum* are yellow turning red at maturity; *D. majus* setae are yellow at maturity.

Similar Species Compared with *D. polysetum*, *D. majus* lacks undulations, has setae yellow at maturity versus red, and is more likely found in northern conifer forests. *D. scoparium*, frequently found with *D. polysetum*, is smaller, not polysetous, and with more organized and tightly packed stems. The photo below shows *D. polysetum* and *D. scoparium* growing together.

Range and Habitat Throughout our range on soil in mixed mesic forests for *D. polysetum*, and from Maine north on soil in damp coastal conifer forests for *D. majus*. See maps next page.

Name *polysetum* for the polysetous condition, and *majus* (L) = great, or large.

Dicranaceae

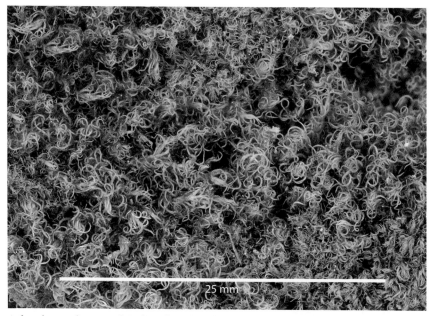

A dry plant on humic soil, Red Trail, Eagle Hill campus, Steuben, ME

Description This small velvety moss is light to dark green, with stems usually in dense tufts to occasionally scattered, to 2 cm long, lower stem parts covered with reddish brown tomentum; leaves, 2–4 mm long, are erect-spreading to falcate-secund when wet, much crisped when dry. Stems may have many branchlets with even smaller leaves. As with *D. flagellare*, *D. fulvum*, and *D. viride*, the capsules produced by this species are straight.

Similar Species See Dicranaceae table. When wet, this species looks much like a lighter green version of *D. flagellare*, but it lacks the flagellate branches and has leaves less tubular and much crisped when dry versus not much changed for *D. flagellare*.

Range and Habitat Widespread in eastern North America from the Canadian Maritimes south in the Appalachian Mountains to North Carolina, on tree bases, tree roots, and rotting wood.

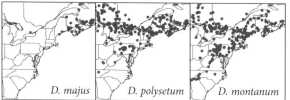

Name *montanum* for the fact that the type specimen was collected in the mountains. This isn't a particularly montane species.

Dicranaceae

Collected in a Downeast conifer forest, Washington County, ME

Description A large *Dicranum* with loosely packed stems to 8 cm tall, densely covered with white to more commonly reddish brown rhizoids; leaves 6–9 mm, somewhat falcate-secund, but messy looking, slightly undulate, with flexuose tips, very contorted when dry; multiple curved capsules per stem.

Similar Species Our other large polysetous *Dicranum* species, *D. polysetum* and *D. majus*, have leaves that contort very little or not at all when dry. *D. undulatum*, also large, has 1 seta per stem, leaves more undulate, and a bog habitat.

Range and Habitat Throughout the northern part of our range in damp to dry conifer forests.

Name Derivation *ontariense* is for Ontario. The former name of this species was *drummondii* for Thomas Drummond, who collected in Ontario. When the name needed to be changed for technical reasons, *D. ontariense* was chosen to commemorate Thomas Drummond in a somewhat distant way.

Dicranaceae

25 mm

On soil in a hardwood forest along the trail to Squirrel Point Lighthouse, Arrowsic, ME

Description This species has tightly packed stems to 6 cm high densely covered with reddish brown to occasionally white rhizoids. The falcate-secund leaves, to 9 mm long, little changed when dry, have small teeth visible with a hand lens near the leaf tips. Sections of the frequently large clumps seem to have the leaf tips bending in the same direction, giving a look of organization to the plant. Setae are initially yellow, turning red with age, capsules are curved. See genus *Dicranum* page for more capsule photos. This is a highly variable species that can vary from tightly tufted as shown above to longer and more silky looking.

Similar Species Most of the *Dicranum* spp. have something of this "broom" or "windblown" look. It is hoped the chart in the Key section will assist in sorting them out. *D. ontariense* is most similar, also with curved capsules, but it crisps when dry and has multiple setae.

Range and Habitat Throughout our range, and much of the world in mesic to damp deciduous or mixed forests on soil or anything with a soil layer over it.

Name *scoparium* (L) = broom, in reference to the look of a worn-out straw broom. See map on previous page.

Dicranaceae

All photos are from Gulf Hagas, ME.

Description Scruffy-looking plants on tree trunks, light yellow-green to dark green, stems to 2 cm long, more or less matted with white to reddish-brown rhizoids below; leaves 4–6 mm long, erect-spreading to incurved or occasionally falcate-secund, not much changed when dry, with many broken tips, presumably as part of a vegetative reproduction strategy. Capsules, not commonly produced, are 1 per stem, and straight, that is, the capsule isn't curved, though it may be inclined.

ID Tip to check for fragile leaf tips, brush a wet fingertip over the colony, and you will pick up some broken leaf tips on your finger. Use a hand lens to check.

Similar Species See Dicranaceae table. The tree trunk habitat, the not particularly falcate-secund leaves, and the broken leaf tips are the key characteristics.

Range and Habitat Throughout our range on tree

D. viride

trunks of various hardwood species plus cedar, but showing a preference for maple bark.

Name *viridis* (L) = green, probably to distinguish this species from the darker *D. fulvum*.

Dicranaceae

Above, herbarium specimen (color unreliable); below, near Squirrel Point Lighthouse, Arrowsic, ME. Drawing by the author.

Description In green to yellow-green tufts to 3 cm high with unbranched stems covered with reddish rhizoids; leaves 4–6 mm long, narrow, subtubulose, squarrose to somewhat secund when wet, very contorted (but not at all appressed) when dry, and with broad, clasping bases. monoicous, and a reliable producer of yellow-brown capsules that are distinctively furrowed, asymmetrical, inclined, and with a pronounced struma (goiterlike swelling). See arrows above. The leaves have distinctive hyaline (possibly whitish when dry) clasping bases, and the capsules have red teeth.

Similar Species See the *Dicranum* chart where this species was included because of its long, narrow leaves, and familial connection. The white leaf bases and distinctive capsules may be diagnostic if fresh specimens show these characteristics. Fortunately, the strumae are easy to see and capsules are reliably produced.

Range and Habitat Throughout our range, south in higher terrain to North Carolina, on rotting logs in cool, damp conifer forests.

Name *onkos* (G) = a protuberance, and *phoros* (G) = bearing, for the swelling at the base of the capsule. *Wahlenbergii* is for Goran Wahlenberg (1780–1851), the Swedish doctor and botanist who collected the type specimen for this species in Norway.

O. wahlenbergii

Dicranaceae

On rock, Saunders Trail, Kennebec Highlands, Mt. Vernon, ME

Description With stems short and unbranched, leaves 4–7.5 mm long and falcate-secund, this looks for all the world like a *Dicranum* sp., but check the *Dicranum fulvum* and *D. scoparium* pages to tease out some differences. The long, narrow, hairlike leaves are little changed when dry, and leaf cells alternate from green photosynthetic cells to clear cells along the length of the leaf, giving it a striate and grayish look. Also note that the costa fills more than half the leaf width at the basal sheath, and essentially all the width above the base. The plant has a gray sheen when dry, but it looks pale green to yellow-green when wet. Straight cylindrical capsules are very Dicranum-like on setae, initially yellow and turning red at maturity.

Similar Species *D. scoparium* has arcuate inclined capsules and is not usually a rock species, though it certainly can be found on soil over rock. See *D. fulvum*, a rock species, for comparison with that species.

10×

Range and Habitat Throughout North America from the Arctic treeline in the north, south to New Mexico in the west, and to South Carolina in the east. Typically found at high elevation on rock, or soil over rock.

Name The alternating green and clear cells are very *Leucobryum*-like, thus the genus name; *longifolium* is for the long leaves.

P. longifolium

2×

Dicranaceae

Above, herbarium specimen; right, along gravelly roadside, Lead Mountain, ME

Description Small, upright, loosely packed stems to 1 cm high; leaves 2–4 mm long, contracted quickly to an extended awn, upright-spreading and erect to flexuose when wet, slightly twisted when dry; seta long and bright yellow; capsule with a distinctive long neck, occasionally strumose (with a goiterlike swelling at the base). Autoicous and a prolific capsule producer.

Similar Species The yellow setae will resemble those of *Ditrichum pallidum*, but these capsules are distinctive. [*Trematodon longicollis*] has leaves gradually tapered to a point, a capsule with a distinctly longer neck—twice the length of the capsule urn versus as long as the urn for *T. ambiguus*. Also, [*Trematodon longicollis*] is likely to be found from NJ south and is found on moist soil, versus dry for *T. ambiguus*. *Dicranella* spp. will have a similar look, but the capsules are quite different.

Range and Habitat Found from Labrador south to NY and Michigan, on relatively dry clay soil in open disturbed sites such as fields, roadsides, streambanks, and gravel pits. Uncommon.

T. ambiguus

Name *trema* (G) = a hole and *odon* (G) = tooth, for microscopic perforations in the peristome teeth, and *ambiguus* for an ambiguous relationship to the *Dicranum* genus as perceived by Johannes Hedwig (see *Hedwigia ciliata*).

Dicranaceae (often in Bruchiaceae)

Gravel driveway edge, Nelson, NH, 0.5×

10×

Description A common moss forming dense irregular velvety mats. Leaves are small, ≈2 mm long, lance-shaped with straight revolute margins from base to tip, costate, and folded up along the costa. Chestnut to red-brown capsules ≈2.5 mm long on purple setae are copiously produced spring through summer. The drawing lower right shows a dry capsule. Spring capsules are shown on the next page. Note the asymmetrical bump at the capsule base.

A. J. Grout

Similar Species The dense reddish swards created by the frequently prolific seta and capsule production, and the keeled leaves with tight revolute margins help separate this from the other waste-place specialists such as *Pohlia, Funaria,* and *Bryum argenteum* with which it frequently shares space. That said, Crum and Anderson (1981) warned that "the sterile forms [of *C. purpureus*] are protean and occasionally fool the most seasoned bryologist." Fortunately, capsules are commonly produced.

Range and Habitat Distributed widely throughout the Northern Hemisphere on gravelly roadsides and generally dry, open, barren soils. An alliance with asphalt, concrete, or charcoal is common.

C. purpureus

Name *cerato* (G) = horned, and *don* (G) = teeth, referring to peristome teeth forked like the horns of some unspecified animal. *purpureus* (L) = purple for the color of the setae and capsules. Common name: purple ceratodon.

A. J. Grout

Ditrichaceae

Continued on next page

Above (2×) and below left (4×), little-traveled dirt road, Hancock, NH; below right, spring capsules with and without calyptra (12×). The slight bulge (a struma, arrows) at one side of the base of the capsules becomes more pronounced as the capsules age and become more curved.

Ditrichaceae

Collected on a soil bank in Arrowsic, ME; capsules shown fresh (right) and dry (below)

Description Small silky plants, green to yellow, to 0.5 cm tall, with short unbranched stems; narrow leaves long-pointed, subulate, falcate-secund, 3–5 mm long, appearing mostly basal, not much changed when dry. Autoicous with frequently produced small-mouthed, long-toothed capsules on bright yellow setae. Mature capsules are curved, and not much, if at all, constricted below the mouth.

Similar Species The commonly produced capsules separate this from the gametophytically similar *Dicranella heteromalla*, which has capsules constricted below an oblique mouth. [*Ditrichum flexicaule*], uncommon in the eastern Unite States but found occasionally in the Canadian Maritimes and the north shore of the Great Lakes, is larger, requires a calcareous rock substrate, has red setae, and its leaves are twisted when dry. [*D. rhynchostegium*], probably also rare but found throughout our range, has orange-red setae. [*Ditrichum lineare*], common in the northern portion of our range, has shorter blunt-tipped leaves lying closely against the stem and is not often found with capsules. *Dicranum flagellare* doesn't have the silky look, produces flagellae, and has very different capsules.

Range and Habitat Throughout eastern North America from Nova Scotia to Florida on bare soil on trail and road banks.

Name *di* (G) = two, and *trich* (G) = hair, for the split filiform peristome teeth. *pallid* (L) = pale, for the blond setae.

D. pallidum

Ditrichaceae

Moist and dry. Left, from a limestone quarry in Rockland, ME; right, on a shell midden, Damariscotta, ME; bottom right, drawings from Bry. Eur.

Description Stems 1–2 cm tall with brownish green filiform gemmae (brood bodies) in the leaf axils. The leaves are entire, 2–4 mm long, with percurrent costae, erect when moist, contorted when dry. Ireland (1982) said sporophytes are rare to unknown on plants from eastern North America. Too bad, since the calyptra is the most distinguishing feature, resembling a candlesnuffer and providing the common name. The *E. procera* capsule is shown on the left compared with the [*E. ciliata*] capsule on the right. Monoicous.

E. procera with brood bodies

Similar Species [*E. ciliata*] plants are smaller and more fertile, producing capsules with a ciliate-margined calyptra.

Range and Habitat Throughout our range, a calciphile on limestone, in calcareous ledge cracks, soil over rock and shell middens. Plants from western North America are reported to be frequently fertile.

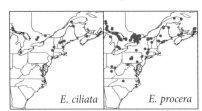

E. ciliata   *E. procera*

Name *encalypta* (G) = veiled, for the fringed calyptra, particularly on *E. ciliata*. *procerus* (L) = tall, possibly for the tall calyptra, or for the height that distinguishes this species from [*E. ciliata*].

Encalyptaceae   Ca

Fissidentaceae (A Single Genus Family)

How we got here: Leaves in two distinct rows, broad, often >1 mm wide, with a flap fused onto the upper surface to form a sheath; on damp, shady, humic soil, rotten logs, and seepy cliffs.

Species characteristics to note: Size of the plant and leaves, whether the seta originates at the base or tip of branch, leaves bordered or not, and as always, habitat. See the chart on the next page for species information.

This clearly delineated family is the exception that proves at least three rules. First is the rule that leaves on all mosses are spirally arranged around the stem, and if they look flat, it's just a matter of appearance—the leaf insertion is really spiral. Well, not so here. These leaves are indeed two-ranked. Second is the rule that any bryophyte with lobed leaves is a liverwort. Again, this family is an exception; *Fissidens* leaves have a connected flap that clasps the stem. Third, acrocarp sporophytes all originate from the tip of a stem. In some members of this family, the sporophyte originates at the base of the branch, but the family is always categorized as acrocarpous.

F. adianthoides showing sporophytes originating from the base. Winslow Ledges, ME

You should be pleased to have made it this far. Reliably arriving at an exact species name, known as the specific epithet, can be quite difficult for this group with or without compound microscope work. The following table and the range maps should give you a start in the right direction.

Species	Notes	Sporo-phyte insertion	Leaf	Size, stem length × width	Habitat
F. adianthoi-des	1,2	Base	Pale border, a few teeth at tip	Large, ≤6 cm × 5 mm	Moist woods and fens
F. bryoides	1	Stem tip	Bordered, entire	Small, <1 cm × <2 mm	Rocks along streams and in forests
F. bushii	1	Base	No border	Small, 0.8 cm × 3 mm	Waste places
F. dubius	1	Base	Pale border, a few teeth at tip	Medium, ≤3 cm × 3 mm	Woods, humic soil, stumps
F. fontanus	1	Short, axillary	No border, entire, long and narrow, soft	Large, ≤12 cm long with 4–8 mm leaves	Submerged in slow-moving water
F. grandi-frons	3	No sporo-phytes	No border, entire, long and narrow, rigid	Large, ≤10 cm, leaves ≤4 mm	Calcareous rocks in quick water
F. osmundi-oides	4	Stem tip	No border, entire	Medium, ≤3 cm × 3 mm	Soil banks and seepy ledges

Notes:
1. Found throughout our range.
2. Can reliably be found in a rich fen.
3. Largely a western species, but reported from the Great Lakes region and in the South from Tennessee and Alabama.
4. Generally a northern species. See range maps next page.

Left, *F. bushii,* Cape May, NJ; above, *F. adianthoides* on limestone outcrop, Winslow Ledge, ME

Fissidentaceae

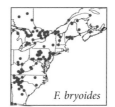

Name *fissi* (L) = split, and *dens* (L) = tooth, for the split peristome teeth of the capsule. Two of the species names, *adianthoides* and *osmundioides*, are for the resemblance to fern genera; *dubius* is just that—I haven't the foggiest; *bryoides* was once thought to resemble a *Bryum* when the concept of that genus was different; *bushii* is for B. F. Bush, who collected the type specimen in Missouri in 1897; *fontanus* is from the Latin word for spring or brook; and *grandifrons* is for its large size and resemblance to fern fronds. Common name: pocket moss.

Fissidentaceae

All photos, including those on the following page, are from the Eagle Hill campus, Steuben, ME; note the spiral teeth shown in the top left inset.

Description As you might guess from the photo above and on the following page, this species is all about the copious capsule cascade. The nodding asymmetrical appearance and long (2–7 cm) twisted setae are distinctive. The leaves are mostly basal, costate, oblong-lanceolate, and 2–4 mm long. When moistened, the seta unwinds and turns the capsules like party balloons. Fortunately this species is a reliable capsule producer.

Similar Species The other moss that lacks obvious leaves and frequents gravelly waste places is *Pogonatum pensilvanicum*, but the capsules of that species are quite different.

Range and Habitat Throughout the nontropical world on bare soil in disturbed places, with a particular preference for recently burned areas. Often found on railroad tracks.

F. hygrometrica

Name *funis* (L) = rope, for the twisted seta, *hygro* (G) = wet, and *metreo* (G) = to be measured in a circular or radiate way, for the tendency of the capsules to wind and unwind with moisture changes. Common name: cord moss.

Funariaceae

Fresh spring capsules and leaves

Funariaceae

Top, on bare soil in Alexandria, VA, well hydrated; above, UMO herbarium collection #2127, collected by Rev. J. Blake in 1864; right, capsules collected 148 years later in Arrowsic, ME

Description One of the few annuals in this book, this species fruits in spring soon after the snow melts and produces abundant, distinctive capsules 1–2 mm long on a yellow-orange seta, ≈1 cm long rising from a stem 3–10 mm tall. Leaves are clustered toward the stem tip and broadly lanceolate, 2–3 mm long, entire to very little toothed, costate, erect-incurved when wet, and lightly crisped when dry. Autoicous, and all about the capsules copiously produced in spring.

Similar Species *Tortula truncata* has leaves and capsules in much the same form, a yellow seta, and entire leaves, but it is much smaller, with stems <0.5 cm tall, and capsules <1 mm long are produced in fall rather than spring for *P. pyriforme*. Also, it is found on calcareous soil versus no soil preference for *P. pyriforme*, and its costa is excurrent versus percurrent, though this is probably hard to see with a hand lens. Fresh capsules are similar to those of *Bartramia pomiformis*, but *B. pomiformis* has long, thin leaves, much longer setae, and larger capsules that are very different when mature.

Range and Habitat Throughout our range on damp bare soil. Often found in gardens in early spring.

Name *physco* (G) = a bladder or blister, and *mitri* (L) = a cap, for the unusual capsule and calyptra; *pyri* (L) = a pear, for the capsule form.

Funariaceae

Grimmiaceae (Rock Mosses)

This family, in our region comprising three major genera, *Grimmia* (p. 147), *Racomitrium* (p. 148), and *Schistidium* (p. 151), is a large and challenging group of mosses.

How we got here: Acrocarpous (mostly see *Racomitrium* discussion below), leaves lanceolate, not long and narrow, twisted or appressed when dry, not contorted, frequently with a strong costa, and white tips or awns; capsules with 16 teeth, cylindrical, not narrowed at neck, frequently immersed in leaves or slightly exserted above the leaf tips; on rock.

Many of these rock moss leaves have hyaline (lacking chlorophyll, appearing white) tips or awns of various lengths, though a few are known without this characteristic.

Those without white leaf tips are referred to as muticous, meaning a white leaf tip does not occur where one would normally be expected. This unusual term is used only with this family.

As with many rock specialists, these plants are often dark green to blackish.

Racomitrium. As you may have figured out by now, the acrocarpous/pleurocarpous distinction is not a clean biological or evolutionary division; it's a useful split to help us key mosses. The *Racomitrium* spp. don't fit neatly into either group. Rather than having the seta rise from the tip of the plant as in most acrocarps, or from the side of the stem as in pleurocarps, the *Racomitrium* plants produce their seta from the tip of a branch. Crum and Anderson (1981) called them pleurocarps. Bruce Allen (2005) used the term cladocarpous to describe this branch-tip fruiting, and Robert Ireland (1987) keyed them as acrocarps. I've included them as acrocarps in part because I'm using Robert Ireland's key, and in part because of the family connection. That said, they do have a generally low, somewhat spread-out growth form rising from horizontal stems, and they do branch (rather than fork), separating them from the rest of the family. *Racomitrium* spp. are all dioicous, and they produce capsules on a seta long enough to get the capsule well above the leaves, whereas the capsules of *Grimmia* spp. are slightly above the leaves, and those of *Schistidium* spp. are immersed in the surrounding leaves.

What most conclusively separates *Racomitrium* from the rest of the Grimmiaceae is its unusual leaf cell structure. *Racomitrium* leaf cells have very wavy walls, referred to as sinuose-nodulose, a very distinctive characteristic under a compound microscope (see right).

The fact that few pleurocarps are found on rock is helpful here. If you get to know *Hedwigia ciliata*, a common and ecostate rock pleurocarp, and can eliminate it, your pleurocarpous-looking rock moss is likely a *Racomitrium*.

Schistidium and *Grimmia:* These are more clearly acrocarpous, erect, growing in tufts, with no or forked branching. In order to tell them apart without microscopic investigation, you need capsules.

Schistidium has enlarged and differentiated perichaetial leaves (the leaves surrounding the capsule), immersed capsules, and a columella connected to and falling off with the operculum. Most species are monoicous and should be good capsule producers.

Grimmia has undifferentiated perichaetial leaves, usually exserted capsules, and a columella that remains in the capsule after the operculum falls off.

Racomitrium lanuginosum in the Chic-Choc Mountains, Québec, Canada

Grimmiaceae

Some of the More Common Species of Grimmiaceae

Racomitrium (see also key next page)

R. aciculare (p. 148): Muticous, blunt-tipped leaves; on wet rocks, occasionally inundated.

R. heterostichum group (p. 149): Hyaline-tipped leaves, on exposed to shaded, dry, noncalcareous rock. Includes *R. venustum*, shown below, which is mostly muticous.

R. lanuginosum (p. 150): The common white-tipped moss of our northeastern alpine zones.

Grimmia

G. muehlenbeckii (p. 147): Small and velvety, with small leaves that are keeled, with margins recurved, and a hyaline awn. Unfortunately, this, our most common *Grimmia* species, is rarely found with capsules.

[*G. donniana*]: alpine, monoicous, prodigous producer of yellow capsules.

Schistidium (see also key next page)

[*S. agassizii*]: Muticous. The only member of the genus with blunt-tipped leaves. According to Bruce Allen (2005), a common riparian species; I find it on asphalt shingle roofs, which is, by the way, a pretty good place to find rock mosses.

S. apocarpum (p. 151): Hyaline-tipped leaves with awn length >1/4 of leaf length, usually on dry rock. Common.

[*S. rivulare*]: Muticous, or with short hyaline awn, <1/4 of leaf length. Riparium species found in or along streams.

S. maritimum (p. 153): Muticous or hyaline-tipped, found in the salt spray zone of our rocky northern coasts.

The mostly muticous *Racomitrium venustum* (see *R. heterostichum* group, page 149), on the vertical face of a roadside rock, Arrowsic, ME

Grimmiaceae

Grimmiaceae Keys

Schistidium

1. Leaves ovate-lanceolate, acute or acuminate, rarely narrowly obtuse, often with a hyaline mucro or awn 2
1. Leaves ligulate to linear-lanceolate, narrowly obtuse to broadly acute, lacking hyaline tip..3

2. Leaves usually with a long awn, 1/4 the length of the leaf or more; dorsal surface of costa often rough near apex; on exposed, usually dry calcareous or noncalcareous rock ....*S. apocarpum*
2. Leaves lacking awn, or awn very short; dorsal surface of costa smooth; on wet rocks along and in streams ....[*S. rivulare*]

3. Leaf apices narrowly obtuse, sometimes acute, leaves not eroded; maritime plants growing on boulders or in cliff crevices in the spray zone beside the ocean..*S. maritimum*
3. Leaf apices broadly obtuse to broadly acute, the leaves often eroded; plants not maritime, on acidic boulders in or beside streams, often near waterfalls ...
   ....[*S. agassizii*]

Racomitrium

1. Leaves with an obtuse, nonhyaline apex; on sandy soil over rock in or beside streams... *R. aciculare*
1. Leaves with an acute, hyaline apex 2

2. Hyaline leaf apices entire or indistinctly toothed; plants yellowish green above, dark green to brown or black below; on sandy soil on mainly acidic boulders and cliffs, often beside streams and lakes .... *R. heterostichum* group
2. Hyaline leaf apices distinctly toothed; plants yellowish green to grayish green above, light brown to blackish below..3

3. Hyaline leaf apices extending down margins as decurrencies, strongly toothed; plants grayish green; on dry exposed acidic soil or rock, frequently alpine..*R. lanuginosum*
3. Hyaline leaf apices not extending down margins as decurrencies, weakly toothed; plants yellowish green to light brown; in exposed habitats on sandy soil over rock or on sand or gravel beside lakes ....
   .... [*R. canescens*] (see *R. lanuginosum*)

Note: No subkey for *Grimmia*. See previous three pages.

Schistidium apocarpum
group showing
immersed capsules

Grimmiaceae

Winslow Ledges, Winslow, ME. Calcareous ledges along the Kennebec River in full sun.

Description Small tuft-forming dark green to blackish-green mosses with unbranched to occasionally forked stems; leaves 1–1.5 mm long, lanceolate, keeled along a strong costa, ending in a hyaline awn. Dioicous, with sporophytes rare, apparently not produced at all within our coverage area. Microscopic reddish gemmae are produced along the leaf bases.

Similar Species *Grimmia* spp. and *Schistidium* spp. are erect-growing, tuft-forming acrocarps with only occasional forked branching, and hyaline leaf tips of varying lengths. The *Schistidium* genus has enlarged and much differentiated perichaetial leaves, immersed capsules, and a columella that is attached to and falls off with the operculum, whereas the *Grimmia* spp. have relatively undifferentiated perichaetial leaves, exserted capsules, and a columella that remains in the capsule when the operculum falls off.

Since this species produces no capsules, it's not possible to distinguish it from genus *Racomitrium* without a compound microscope. See *Schistidium* and *Racomitrium*.

Range and Habitat A northern species, from Labrador south through New England and west to Wisconsin, on stone walls and on rocks along water.

Name *Grimmia* is for Johann Friedrich Karl Grimm (1737–1821), amateur German botanist; *muehlenbeckii* is for Heinrich Gustav Muehlenbeck (1798–1845), Alsatian physician who studied the flora of Alsace.

G. muehlenbeckii

Grimmiaceae

On an occasionally inundated streamside rock, Camden State Park, Camden, ME

Description Large plants with stems 3–8 cm long; leaves 2–2.5 mm long, wide spreading when wet, to appressed when dry, blunt-tipped, costate, entire, and without the hyaline tip so common in this genus. Dioicous, as are all species in this genus, with capsules ≈2 mm long. The setae in this species are twisted clockwise when viewed from above.

7×

Similar Species *R. aduncoides* is a new species recently separated from *R. aciculare*. The new species has a shorter costa and narrower leaves, and it favors drier habitats. It's probably rare.

Range and Habitat A common moss on wet rocks along streams or under waterfalls.

Name *rhakinos* (G) = ragged, and *mitra* (G) = hat, for the ragged margin on the calyptra (see above right); *acicula* (L) = a small point for the capsule tip.

10×

R. aciculare

Grimmiaceae

5 mm

R. heterostichum var. *microcarpum*, Eagle Hill campus, Steuben, ME

Description *R. heterostichum* var. *microcarpum* (shown above) is a typical member of the *R. heterostichum* group, which includes *R. affine, R. sudeticum, R. venustum,* and *R. heterostichum* var. *microcarpum,* all listed as *R. heterostichum* varieties by some authors. They are dark green to yellow above, reddish brown to black below, with stems to 5 cm long; leaves 2–3 mm long, entire, lanceolate to obtuse, with recurved margins and ending in a hyaline tip of varying lengths. Dioicous, as are all species in this genus, with capsules ≈2 mm long, and setae twisted clockwise when viewed from above.

Similar Species Most of the plants will key to *R. heterostichum* var. *microcarpon,* or in some texts, *R. microcarpon,* or possibly *R. venustum,* which has leaves curved into a hooklike branch tip and is frequently muticous (see *Racomitrium* genus page). *R. affine* is rare and very northern, and *R. sudeticum* is a rare alpine species. The key to separating these species is unfortunately quite technical, and is based on details of the alar cell and the extent of cellular papillae, among other things. Note that the hyaline leaf tip in this group begins rather abruptly compared with that of *R. lanuginosum,* where the hyaline portion runs down the side of the leaf in an inverted V pattern. You can either go back to that compound scope, or try a 20× hand lens, or just be happy you've gotten this far.

Range and Habitat Throughout the northern part of our range, south to Michigan and NY, possibly to North Carolina in the mountains. On exposed to shaded, usually dry, noncalcareous rock.

Name *hetero* (G) = other, and *stich* (G) = row, perhaps indicating that the leaves are in opposite rows, but that is not apparent.

R. heteros-tichum

Typical leaf for this group, 20×

Grimmiaceae

Racomitrium lanuginosum in the alpine zone of Mont Albert, Chic-Choc Mountains, Québec, Canada. See additional photo on page 144.

Description Plants in frequently extensive colonies with a white, woolly look caused by the long hyaline leaf tips. Stems are to 12 cm long with leaves 3–5 mm, lanceolate, margins recurved, and with long white papillose hair points that extend down the sides of the leaves forming an inverted V, as shown below. Dioicous, capsules rare.

Similar Species The very long hyaline leaf tip and the alpine habit help separate this species from those on the previous pages. Note that the white tip extends down the sides of the leaves, unlike the hyaline tips of *R. heterostichum* group. [*R. canescens*] also has long hyaline hair points, but they begin more abruptly than the tips of this species, and the costa extends only 1/2 to 2/3 of the leaf length versus to the tip of the nonhyaline portion in *R. lanuginosum*.

5×

Range and Habitat This is the common long, white-tipped moss of our northeastern alpine zones. On exposed soil or rock.

R. lanuginosum

20× (Bry. Eur.)

Name *lanuginosus* (L) = woolly, for the white-woolly look of the frequently extensive colonies. Common name: woolly moss.

Grimmiaceae

Above, on dry granitic rock in Nelson, NH; right, on calcareous rock along the Kennebec River, Winslow, ME

Description Small, dark green plant with densely packed stems to 3 cm long; leaves 2–3 mm long, lanceolate, erect to erect-spreading when wet, to appressed and somewhat twisted when dry, keeled, margin recurved, and with a hyaline awn of varying length depending on species. Monoicous, capsules immersed, with dark red teeth and red operculum.

Similar Species *Grimmia* spp. and *Schistidium* spp. are erect-growing, tuft-forming acrocarps with only occasional forked branching, and hyaline leaf tips of varying lengths. The *Schistidium* genus has enlarged perichaetial leaves, immersed capsules, and a columella that is attached to and falls off with the operculum, whereas the *Grimmia* spp. have

Capsule surrounded by mostly muticous leaves

relatively undifferentiated perichaetial leaves, exserted capsules, and a columella that remains in the capsule when the operculum falls off. *S. rivulare* is a very similar species found on rock along streams or rivers. It lacks hyaline leaf tips and is easily confused with the occasional muticous population of *S. apocarpum* (see capsule photo). See family page for more information.

S. apocarpum

Range and Habitat Common throughout our coverage area, south to Georgia on dry rock, in shade to partial sun. *S. rivulare* is found on occasionally inundated rock.

Grimmiaceae   Continued on next page

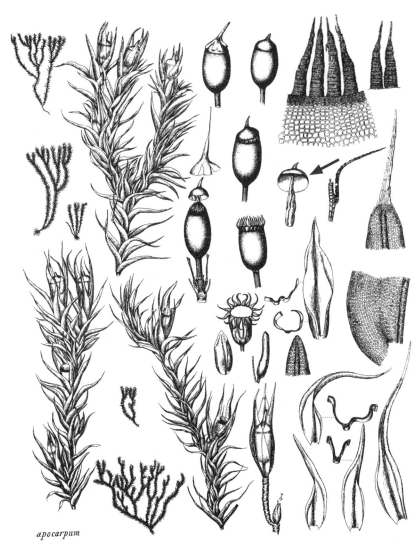

apocarpum

In *Schistidium* spp., the columella, the inside of the capsule, stays attached to the operculum when it comes off (red arrow); in the other Grimmiaceae, the columella stays with the capsule when the operculum is shed. Scale is quite variable here, but the plants top left are approximately life-size, and the capsules are usually ≈1 mm long, making the drawings above approximately 10×. Bry. Eur.

Name *schistos* (G) = split, and *idium* (G) = diminutive, for a calyptra that has a split base; *apo* (G) = off or away, and *carpum* (G) = a fruit, for the way the columella tears away from the rest of the capsule with the operculum.

Grimmiaceae

In rock crevice at the wrack line, Indian Point, Georgetown, ME. Note the grains of sand in the lower right photo.

Description Dark green moss in dense velvety tufts to 2 cm high (usually less); leaves 2–3 mm long (perichaetial leaves seen above are ≈4 mm long), costate, lanceolate, concave, with recurved (usually) or flat margins, erect-spreading when wet, not much changed when dry, without the hyaline tips common on many rock mosses. Because this plant is monoicous, we might expect it to be a good capsule producer, but that is certainly not always the case. The capsule photo above required much hunting. When produced, the capsules are immersed and have photogenic red teeth. As you can see above left, this population, capsules and all, is only slightly above the high tide line.

Similar Species Many of the *Grimmia* and *Schistidium* species form dark green, velvety tufts, but this is the only moss in the Grimmiaceae family found in the salt spray zone. See *S. apocarpum* and Grimmiaceae pages.

Range and Habitat From the Canadian Maritimes south to Massachusetts on acidic coastal rocks in the salt spray zone.

Name *maritimum* is for the maritime habitat. Common name: salt-spray moss.

Grimmiaceae

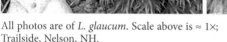

All photos are of *L. glaucum.* Scale above is ≈ 1×;
Trailside, Nelson, NH.

5×

Description A very common trailside moss growing
in small tufts <1 cm high in the case of *L. albidum,* to
large tufts up to 12 cm high for *L. glaucum.* Leaves of
both species are whitish because of the presence of
some cells lacking chlorophyll, but color varies
from green to quite bleached-looking, as shown
above. Leaves are subtubulose, 2–4 mm long in *L.
albidum,* and 3–9 mm long in *L. glaucum.*

Similar Species The two species can be difficult
to distinguish with smaller specimens where
their ranges overlap. If leaf length and cushion
size don't separate them, you may have to resort to
a microscopic analysis, comparing the relative lengths
of the narrow leaf tips to the leaf bases. Capsules may be

Capsule cluster

a help in that *L. albidum* is reported to be more likely to fruit than *L. glaucum,*
though fruits are not uncommon in either species. Technical references give
seta length as 8–12 mm for *L. albidum* and 8–18 mm for *L. glaucum,* so <12 mm
is a toss-up, and >12 mm is probably *L. glaucum.* Both are dioicous. See family
description (p. 93).

Range and Habitat *L. glaucum* is widespread throughout our range, while *L.
albidum* has more of a southeastern U.S.
distribution, occasionally making it as far
north as southern and coastal Maine.

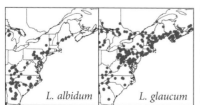

L. albidum   L. glaucum

Name *leuk* (G) = white, and *bryum* (G) =
moss, for a white moss; *glaukos* (G) =
silvery, and *albus* (L) = white. Common
name: cushion moss.

Leucobryaceae

Mniaceae

We discuss here four genera included in the family Mniaceae: *Mnium*, *Plagiomnium*, *Pseudobryum*, and *Rhizomnium*.

How we got here: These are medium to large upright plants with nodding to horizontal capsules, and large, costate, ovate to obovate to broadly lanceolate leaves with a (usually) differentiated border. *Pseudobryum cinclidioides*, lacking the differentiated border, is the exception.

A little page flipping will give you a feel for this family, and you can think of them as *Rhizomnium*, the untoothed Mniaceae; *Plagiomnium*, the single-toothed Mniaceae; and *Mnium*, the double-toothed member of this group. *Pseudobryum* has large ovate leaves lacking a border or teeth.

Rhizomnium spp. key as large acrocarpous plants with obovate to ovate costate leaves with a differentiated, frequently reddish, untoothed border. Capsules cylindrical. All species are found on wet soil or humus, primarily in swampy forests but also on streambanks and margins of lakes.

Rhizomnium

A. Plants medium-large, stems seldom reaching 5 cm high and leaves 3–6 mm long; costae subpercurrent; rhizoids restricted to leaf axils in lower part of stem...*R. punctatum*
A. Plants large, stems often >5 cm high and leaves 6–11 mm long; costae often percurrent; rhizoids scattered along lower parts of stems .... B

B. Leaves with strong reddish border ....*R. appalachianum*
B. Leaves lacking strong reddish border ....*R. magnifolium*

Mnium and ***Plagiomnium*** spp. differ from *Rhizomnium* by the presence of toothed, lanceolate to ovate-lanceolate leaves, and most species have light-colored differentiated borders.

1. Leaf borders with at least a few teeth in pairs (can be challenging to see with a hand lens); costae red, especially near base of leaves; stoloniferous shoots absent...*Mnium*

Mnium

A. Leaf margins without differentiated border of cells, singly serrate or a few teeth in pairs; on soil over rock, sometimes limestone, on banks and on bases of trees...[*M. stellare*]
A. Leaf margins with a differentiated border of cells, doubly serrate.............. B

B. Costae toothed on dorsal surface C
B. Costae smooth on dorsal surface... D

C. On soil on bluffs and cliffs, often on sandstone and limestone, and on bases of conifers..[*M. ambiguum*]
C. On damp, shaded soil banks and streamsides, common.............*M. hornum*

D. Leaves scarcely reaching 1.5 mm wide; plants seldom with sporophytes; peristome teeth yellow to light brown; on soil, frequently over calcareous bluffs and cliffs, sometimes on humus and bases of trees [*M. marginatum*]

D. Leaves often >1.5 mm wide and up to 3 mm wide; plants usually with sporophytes; peristome teeth red to purplish brown .... *M. spinulosum*

1. Leaf borders with single teeth; costae yellow or green; stoloniferous shoots present ....*Plagiomnium*

Plagiomnium

A. Leaves obovate; margins toothed to middle.... *P. cuspidatum* (see *P. ciliare*)

A. Leaves elliptic; margins toothed nearly to base .... B

B. Plants with sporophytes solitary in the perichaetia; occurring in moderately dry habitats .... *P. ciliare*

B. Plants with sporophytes clustered (1–4) in the perichaetia; occurring in swampy or wet habitats on soil, humus, rocks, often in wet depressions in woods.. [*P. medium*] (see *P. ciliare*)

Pseudobryum cinclidioides (p. 161) lacks teeth and a differentiated border, but has ovate leaves and shares much of the look and feel of the family.

A Mniaceae Sampler

Rhizomnium punctatum

Plagiomnium cuspidatum

Mnium hornum, dry

Plagiomnium cuspidatum capsule, dry

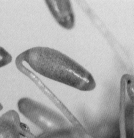
Plagiomnium cuspidatum capsule, fresh

Rhizomnium magnifolium

Zak Preserve, Boothbay Harbor, ME

Description Forms large, light to dark green cushions of overlapping stems to 6 cm long, often complanate, minimally branched, green at tip, ridged from leaf decurrencies, dark brown to reddish and covered with reddish brown tomentum below; leaves 3–5 mm long, with paired teeth (hard to see with hand lens), a differentiated margin, a strong costa that stops short of the leaf tip and is toothed on the back, and dark on lower leaves. The leaves twist mildly when dry, as shown. Plants with antheridial cups as shown above (inset) are common and have surrounding leaves more oblong, and with costa reaching the leaf tip; stems upright, not overlapping. Capsules have 16 yellow teeth and a cone-shaped operculum. Dioicous.

5+

Similar Species *Mnium* spp. leaves have double-toothed margins (see illustration, next page), *Plagiomnium* spp. leaves have single teeth, and *Rhizomnium* spp. leaves are entire. *Atrichum crispum* 5 × leaves may also have double teeth, a similar leaf shape and habitat, but it has lamellae (few, low, and hard to see), its costa extends to the leaf tip, its capsules are different, it's not complanate, and it's much less common. See *Atrichum crispum* for comparison (p. 174).

Range and Habitat Throughout our range on damp shaded soil banks and streamsides, always near water.

Name *mnium* (G) = moss, and *horion* (G) = boundary, for the leaf border.

M. hornum

Mniaceae

All images are of herbarium specimens. Hash marks along the bottom of the top photo are 1 mm apart. Drawing by Mary V. Thayer, in Grout 1903.

Description Forming erect, loose tufts, dark green and sometimes with a red tint, to 2.5 cm high, usually smaller, with branched to occasionally unbranched stems, yellowish green near the stem tips, dark reddish and tomentose below. Leaves crowded toward the tips, 2.5–4 mm long, larger in the terminal cluster than along the stems, with a well-developed double-toothed margin, and a percurrent costa. Leaf margin and costa are often reddish as shown here. Leaves are erect-spreading to spreading when wet, and only slightly contorted when dry. Monoicous and a reliable producer of distinctive red-toothed capsules.

Similar Species *M. hornum* has narrower, longer leaves, is dioicous, and has yellow-toothed capsules. *Atrichum crispum* may also have double teeth, but it has lamellae (tough to see sometimes), and it doesn't have these lovely capsules.

Range and Habitat Throughout the northern part of our range, south to NY and Michigan on soil, humus, soil over rock, tree bases, and rotting logs in forests, possibly preferring conifer or *Thuja* forests. Uncommon.

Name *spinulose* (L) = with thorns or spines, for the well-developed marginal teeth. A common name is flapper mouth, for the flashy red mouth.

M. spinulosum

Mniaceae

The three photos are all of *P. cuspidatum*, Arrowsic, ME; top left, leaves clustered at the tips of fertile stems; top right, tapering vegetative branches (dry); bottom right, capsule cluster. Drawing by Mary V. Thayer, in Grout 1903.

P. ciliare   *P. cuspidatum*

Description The *Plagiomnium* spp. are the single-toothed members of the three genera that were once all placed in the *Mnium* genus. The current *Mnium* genus includes plants with double-toothed leaves, and *Rhizomnium* species lack teeth.

P. ciliare plants are 1.5 cm high, growing from sparsely leaved stolons that can be 10 cm long; leaves ovate, 4–6 mm long, spreading when wet, crisped when dry, toothed all the way to the base with long teeth.

P. cuspidatum plants are 2–3 cm tall, growing from sparsely leaved stolons to 3 cm long; leaves obovate, 1.5–4 mm long, spreading when wet, crisped when dry, toothed only in the upper half.

Both plants can have densely tomentose stems and leaves smaller toward the stem tips, except in the case of fertile stems, which have leaves larger and more crowded toward the tip. Leaves of both species are bordered with lighter-colored cells and may have cuspidate tips. In the case of *P. cuspidatum*, the cuspidate tip is caused by the excurrent costa not found in *P. ciliare*.

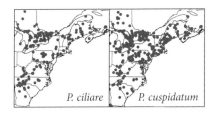

P. ciliare   *P. cuspidatum*

Mniaceae   Continued on next page

Separating the Cousins *P. cuspidatum* is monoicous, produces a single seta per stem tip, and is much more likely to fruit than the diocous *P. ciliare*, which also produces single setae. This, the leaf size, and leaf toothing are the best ways to tell these plants apart. *P. medium* (not illustrated) is also found throughout our range. It produces multiple setae per stem tip, is monoicous and likely to fruit, but is more common in swampy places or in wet depressions in forests and has wider leaves toothed all the away to the leaf base.

Range and Habitat Both species are common throughout our range on soil in shaded mesic forests, or occasionally on tree bases, or on soil over rock.

Name *plagio* (G) = sides or flanks, for the horizontal stolons, and *mnium,* for the genus that once housed these species; *cuspidatum* refers to the excurrent costa of that species, and *ciliare* refers to that species' larger teeth.

Mniaceae

Channing Trail, Harris Center for Conservation Education, Hancock, NH

Description Green to yellowish-green plants in loose mats with stem length ranging from the 4 cm shown in the specimen above to as much as 3× that; stems ridged, green at tips, brown, and with brown tomentum below; leaves distant, oval, 5–7 mm long, costate, entire, without border (microscopic border possible), ending frequently in a slight mucro. The large cells form a pattern visible with a hand lens that radiates up and away from the costa in diagonal rows. Capsules are rare; dioicous.

Similar Species *Mnium*, *Plagiomnium*, and *Rhizomnium* spp. can also show this radiating cell pattern, but their leaves are either toothed to varying degrees, or have a distinct margin visible with a hand lens, or both. *Bryum* spp. are much smaller, with leaves usually bordered and clustered toward the tips, or at least not distant as above.

Range and Habitat Throughout the northern parts of our range from eastern Canada south through PA. Rare as far south as VA. In woods in damp mud depressions. Grout (1903) indicates that this is a rare species in the Northeast, but it's well represented in herbaria (ME, MO, NY), and Ireland (1982) reported it common in the Canadian Maritimes.

Name *pseudo* (G) = false, combined with the genus name *Bryum*, for the similarity in leaf shape to the *Bryum* spp.

P. cinclidioides

Cinclidium for a genus with that name, and *-oid* (G) = looks like, for its similarity to that genus not found in our region.

　　Note: This species will appear as *Mnium cinclidioides* in many references.

Bry. Eur.

Mniaceae

On mucky soil in a wet drainage in a cedar swamp near Ledge Falls campsite on the St. John River, Aroostook County, ME. The dark central dot is the antheridial head (sometimes called a splash cup) where swimming sperm are produced in anticipation of rain allowing the sperm to seek female plants.

Description A large, dark to reddish-green moss with stems to 10 cm long rising from a dense rhizoidal mat; leaves large, 6–11 mm long, entire, very broad, with a strong dark margin, a costa extending to the leaf tip, and tomentum on the decurrent leaf bases and stems. Dioicous, with seta 14–35 mm long, yellow turning red, and the capsule, as shown, is short-necked and cylindrical, to 3.5 mm long.

Similar Species See next page. The other common *Rhizomnium* species, *R. punctatum,* has a strong border like this, but its leaves are smaller. [*R. magnifolium*] also has large leaves, 6–11 mm long, but it lacks the strong, deeply colored border and costa, and it's primarily a northwestern species, less common in the Northeast.

Range and Habitat Throughout the northern portion of our range, south to Pennsylvania with an apparently disjunct population in western North Carolina; in very wet habitats, usually in mucky areas along streams or intermittent drainages.

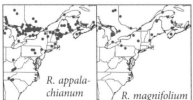

R. appala-
chianum

R. magnifolium

Name *rhiza* (G) = root combined with *Mnium* for a moss rising from a mat of rhizoids. *Appalachianum* is a bit of a mystery. The type specimen for this species is from Ontario, Canada. *Magnifolium* for the large leaves.

Mniacea

Rhizomnium punctatum (reddish) mixed with *Plagiomnium ciliare* (green), *Atrichum undulatum* (yellow-green), and a cork from *Beaujoulais* sp.

Description A medium-large, dark to reddish-green moss with stems to 5 cm long rising from a dense rhizoidal mat; leaves large, 3–6 mm long, entire, very broad, with a strong, frequently red, margin and costa. The costa stops at, or just short of, the leaf tip, which may or may not have a small mucro. The stems are densely covered with rhizoids. Antheridial heads (see photo below right) are common. Dioicous, with capsules as described and illustrated for *R. appalachianum*.

Similar Species *R. appalachianum* (previous page) has a strong reddish border like this, but it has much larger leaves, 6–11 mm long, and a more northerly distribution. [*R. magnifolium*] also has large leaves, 6–11 mm long, but it lacks the strong deeply colored border and is primarily a northwestern species, uncommon in the Northeast.

R. punctatum

Range and Habitat *Rhizomnium* spp. are found throughout our range in very wet habitats; usually in mucky areas along streams or intermittent drainages.

Name *punctatus* (L) = spotted, for the dotted appearance of the antheridial heads.

Mniaceae

Orthotrichum ohioense, on bark in rich woods near Cherryfield, ME

Description These species are typically dark, growing
in tufts on trees, with leaves wide-spreading when
wet, folded upward and appressed when dry,
strongly costate, margins usually recurved and
therefore appearing dark; mature capsules on
short stalks, furrowed with 8 or 16 ridges, and
16 teeth (actually 8 teeth split in half) visible
around the open mouth.

Monoicous, and frequent capsule producers.
The calyptrae are very hairy, though usually less
so than *Ulota* spp., which have leaves much
contorted when dry.

Common Species All on tree trunks except *O. anomalum*, which is found on
calcareous rock. Except as noted, leaves lanceolate and with recurved margins.
Capsule detail is for dry capsules. Most species favor relatively open forest sites.

O. anomalum (see p. 167): Capsules above surrounding leaves, with 8 long
ridges alternating with 8 short ones. Limestone. *Ulota hutchinsiae* is similar,
on noncalcareous rock.

O. obtusifolium: Broadly obtuse leaves with plane to revolute margins.
Capsules 8-ribbed and partly immersed in surrounding leaves. Possibly a
poplar specialist. A look under a microscope (100×) should show 4–5 brood
bodies on the upper leaf surfaces.

O. ohioense: Capsules immersed, not contracted below mouth, 8-ribbed.

O. sordidum: Capsules immersed to slightly exserted, strongly contracted
under mouth, expanded under contraction to a diameter larger than the
mouth. Orange reflexed teeth.

Orthotrichaceae   Ca in part

O. speciosum: Capsules slightly exserted, smooth to obscurely ribbed (8, near mouth), not much contracted under the mouth. Uncommon and northern.

O. stellatum: Capsules immersed to slightly exserted, 8-ribbed, strongly contracted below mouth, and narrow below constriction, less than or equal to the diameter of the mouth.

Range All may be found throughout our range except *O. sordidum*, which is not found south of NY and Michigan, and *O. speciosum*, another northern species.

O. anomalum

O. obtusifolium

O. ohioense

O. sordidum

O. speciosum

O. stellatum

Orthotrichum anomalum capsules

Orthotrichaceae

Above, on limestone, Ravena, NY; inset, capsule covered by its hairy calyptra; below, a population growing on mortar in a rock wall outside the Curtis Memorial Library, Brunswick, ME

Description A dark, scraggly-looking rock moss in sparse to dense cushions; leaves lanceolate, and with recurved margins. Capsules are above surrounding leaves; they are shown above with the hairy calyptra typical of this family. When dry, the capsule has 8 long ridges alternating with 8 short ones. See previous page for capsule image.

Similar Species *Ulota hutchinsiae*, a close relative, is similar, on noncalcareous rock. Mature capsules may help separate the two species. *Ulota hutchinsiae* capsule teeth reflex when dry versus not on this species, and the *Ulota* lacks the alternating long and short ribs. With fresh capsules, note that the calyptra hairs completely cover the *O. anomalum* capsules, whereas with *U. hutchinsiae* capsules (p. 169), the bottom 1/3 is bare, giving a sort of "high and tight haircut" look.

Range and Habitat Throughout our range on limestone, other calcareous rock, and mortar in man-made structures.

Name *Ortho* (G) = straight, and *trich* (G) = hairs, for the calyptrae with straight, upright hairs typical of the family; *anomalus* (L) = abnormal, in recognition of its rock substrate in this largely tree-dwelling family.

Orthotrichaceae   Ca

Above, *Ulota crispa*; inset, a typical tuft, Arrowsic, ME

Description Typically in neat tufts from about the size of a nickel to larger than a quarter, though occasionally somewhat spread out and less neat. The leaves are erect, spreading when wet to very contorted and twisted when dry in *U. crispa* (above), and little contorted when dry in *U. coarctata* (right) and *U. hutchinsiae* (see next page). The two species illustrated on this page are our common tree trunk species. Both are frequent capsule producers, and they are best separated by seta length and capsule shape: short seta and capsules cylindrical and wide-mouthed for *U. crispa,* and longer seta and capsules pear-shaped and small-mouthed for *U. coarctata. U. hutchinsiae* is a rock species with capsules much like those of *U. crispa.* All have hairy calyptrae (as shown).

Ulota coarctata

Ulota coarctata

Similar Species [*U. phyllantha*] is a rare coastal species found from Maine through the Canadian Maritimes. It never produces capsules in our region and is found primarily on rock, sometimes on tree trunks and branches in the salt spray zone, produces clusters of red to red-brown vegetative propagules clustered around the leaf tips, and otherwise resembles *U. coarctata. Orthotrichum anomalum*, similar to *U. hutchinsiae*, favors limestone.

Range and Habitat Throughout our range south to Georgia and Tennessee on substrate discussed above.

Orthotrichaceae

Continued on next page

Ulota hutchinsiae. On rock, Black Mountain Trail, Hancock County, ME. The leaves of this species are not crisped and have a strong costa and recurved margins. Note how the calyptra hairs extend upward. This is typical for the family Orthotrichaceae, while calyptra hairs in the Polytrichaceae family grow downward.

Name From *oulos* (G), meaning curly or twisted; *crispa* (L) = curled; *co-arctatus* (L) = drawn together, for the capsule mouth; *hutchinsiae* is for Ellen Hutchins (1785–1815), Irish bryologist.

U. coarctata

U. crispa

U. hutchinsiae

Orthotrichaceae

Polytrichaceae (Hair-Cap Mosses)

The hair-caps are named for the tomentose caps (the calyptrae) that cover the spore capsules in most of the genera in this family (see *Polytrichum commune* page). In our area the family is represented by four genera: *Polytrichum, Atrichum, Pogonatum,* and *Polytrichastrum*. As you might guess from the name, the *Atrichum* genus belies its familial name by having sparse or no hairs on the calyptra; the other three genera all have appropriately hairy calyptrae.

All the hair-caps share an upright growth form with little or no branching, leaves with sheathing bases and blades growing outward from the main stem in all directions, and lamellae, long rows or walls of cells several cells high by 1 cell wide running parallel to the length of the leaves. Depending on genus, there can be anywhere from very few rows of lamellae covering <1/8 of the leaf width (*Atrichum*) to many rows of lamellae obscuring the entire width of the leaf. The hair-caps have a primitive vascular network and dry fairly quickly after being picked—a trait that helps separate the *Atrichum* genus from the rest, as you'll see in the following key.

Polytrichaceae Master Key

1. Capsule ridged and either square or 6-sided in cross section; calyptra densely hairy; dry leaves folding up along the stem (see *P. juniperinum*), not contorted; moist leaves not undulate........see *Polytrichum/Polytrichastrum* key
1. Capsule cylindrical, with or without dense hairs on the calyptra; dry leaves contorted or not...2

2. Calyptra without hairs (may be a few at the tip); leaves contorted when dry, spreading when moist; lamellae <10 covering 1/3 or less of the leaf lamina.. see *Atrichum* key
2. Calyptra with dense hairs; dry leaves more or less folded up along the stem, incurved and not much contorted, moist leaves appressed to spreading, frequently clustered toward the tip; many lamellae covering most or all of the leaf lamina..3

3. Leaves >6 mm long, <1 mm wide .... *Polytrichastrum alpinum*
3. Leaves <6 mm long, >1 mm wide .... see *Pogonatum* key

Polytrichum/Polytrichastrum

1. Leaves with no teeth and lamellae covered by infolded margins (illustration a, p. 171)...2
1. Leaves with visible teeth (10×), and lamellae not covered by infolded margins (illustration b)..4

2. Tall plants (frequently >4 cm) growing with *Sphagnum* spp. in a bog, leaves generally spreading at the tip, more erect below, stem covered with whitish rhizoids .... *Polytrichum strictum*

2. Not as above 3

3. Short plants (usually <2 cm high) with long white awns (illustration c) ....*Polytrichum piliferum*

3. Short or tall plants with reddish leaf tips .... *Polytrichum juniperinum*

4. Large plants with stocky capsules <1.5× as long as wide (illustration d) .... *Polytrichum commune*

4. Large plants with ridged capsules >1.5× as long as wide (illustration e).........5

5. Plants of northern or high-elevation conifer forests ....*Polytrichastrum pallidisetum*

5. Plants throughout our range in deciduous forests...... *Polytrichastrum ohioense*

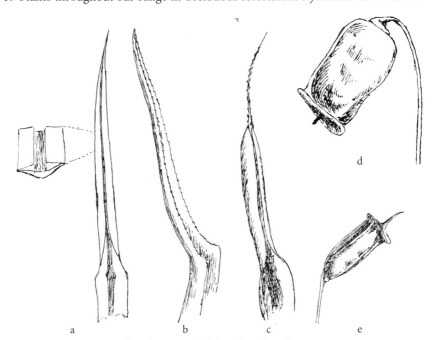

Key characters in *Polytrichum*. Bry. Eur.

Atrichum Key

1. Lamellae cover 1/4–1/3 of midleaf lamina.....................................*A. angustatum*
1. Lamellae cover <1/8–1/10 of midleaf lamina..2

2. Hydrated leaves undulate and with teeth on back of undulations (may require 20× lens or young eyes)...*A. undulatum*
2. Hydrated leaves neither undulate nor toothed on back of lamina. 3

3. Leaves toothed to below middle..*A. crispum*
3. As above, but leaves well toothed only in the upper 1/4................[*A. tenellum*]

Pogonatum Key

1. Leaves inconspicuous, setae and capsules appear to be growing from green soil. 2
1. Leaves conspicuous, toothed, and fleshy-appearing because of many lamellae covering upper surface of leaf...3

2. Plants scattered; leaves toothed; throughout our range............*P. pensilvanicum*
2. Plants more crowded; leaves entire; NY south..
....*P. brachyphyllum* (see *P. pensilvanicum*)

3. Stems in dense clusters, occasionally branched with leaves clustered toward stem tips, alpine..*P. urnigerum*
3. Stems in loose tufts or scattered individuals, mostly unbranched, with leaves well distributed along the stem, alpine or not....................................*P. dentatum*

Above, Eagle Hill, Steuben, ME; below, Mt. Vernon, ME

Description Small plants 1–2 cm high with weakly undulate, toothed leaves <1 mm wide, and with high, very visible lamellae covering ≈1/4–1/3 of the leaf width at midleaf, more nearer the tip. The bottom right photo shows the antheridial cups frequently encountered in this family. These cups, called splash cups, hold sperm that are ejected when a raindrop lands in this very specialized structure.

Similar Species *A. undulatum* and its variations are typically found in wetter sites; they are generally larger, more strongly undulate, and have lamellae covering 1/8–1/10 of the leaf width. *A. crispum* is not undulate at all and has very narrow lamellae. In the rest of the family, lamellae completely cover the leaf lamina, or the leaf lamina is folded over and hides the lamellae.

Range and Habitat Throughout our range on exposed, frequently disturbed, sandy soil.

Antheridial cups, 10 ×.

A. angustatum

Name *a* (G) = without, *trich* (G) = hairs, referring to the lack of hairs on the calyptra in a family in which hair-capped capsules are otherwise the rule; *angust* (L) = narrow, for the narrow leaves.

Polytrichaceae

Above, plants collected from mucky soil at Reid State Park, Georgetown, ME; inset, dry *Atrichum crispum* (left) compared with dry *Mnium hornum* (right); middle right, capsule collected by Joanne Sharpe on Allen Island, Knox County, ME; lower right, capsule from herbarium specimen

Description Plants 1–5 cm high with green to reddish stems and broad leaves <4× as long as wide, not at all undulate when hydrated, without teeth on the back of the leaf lamina (see *A. undulatum* for comparison), with low, indistinct lamellae covering <1/8 of the leaf width. Leaves are toothed all around with single to infrequently double teeth, becoming very contorted (crisped) when dry. Capsules are wide-mouthed, have 32 teeth, and are uncommon.

Similar Species When dry, it looks like a small *A. undulatum,* but with wide-mouth capsules. *Mnium hornum* looks very similar and may be difficult to separate without capsules, though *M. hornum* has leaves less crisped when dry and is mildly complanate. See comparison photo above. *M. hornum* capsules have a cone-shaped operculum and 16 teeth versus a long-snouted operculum and 32 teeth for this species. Also, the *Mnium* may have reddish-brown tomentum on the stem, while the *Atrichum* does not, and the *Mnium* has a dark stem, ridged from leaf costa decurrencies.

Range and Habitat Grout (1903) said it's rare. Ireland (1982) reported it occurring frequently throughout the Maritimes and south to Florida. Crum and Anderson (1981) said it's uncommon. I think it's undercollected because of its similarity to *Mnium hornum*. Habitat is shaded, wet, sandy or peaty soil along brooks, or on banks of roadside ditches.

A. crispum

Name *crisp* (L) = curled or wrinkled.

Var. *oerstedianum*; above, Eagle Hill, Steuben, ME

Description Generally large plants but can range from 1 to 6 cm tall depending on expression (see below). Leaves narrow, >5× as long as wide, undulate when moist, crisped and contorted when dry, with lamellae few, covering <1/8 of the midleaf width, and with small teeth on the margins and on the backs of the undulations.

2 ×

Similar Species The abbreviation s.l. following the species name stands for *sensu lato*, meaning in the broad sense, indicating that several possible species are included in this large tent. The various expressions of this species, *undulatum*, *altecristatum*, and *crispulum* are frequently given species status. Crum and Anderson (1981) said, "the distinctions are often intangible or at least difficult to establish and intergradations are puzzling. . . . Precise determination of fertile and fruiting material is difficult, and determination of sterile material is close to impossible. We therefore reserve the option of using the name *A. undulatum* in a broad sense while allowing for the identification of good material as varieties." For the purpose of this book, the very conservative species concept of Crum and Anderson seems reasonable.

2 ×

3 ×

A. undulatum

Range and Habitat All three varieties (or species) are common throughout the Northeast on soil in shaded and damp situations, frequently along streams or roadside ditches.

Name *undulatum* for the undulate leaves. The common name, wavy *Catharinea*, is because this genus was once named *Catharinea*.

Polytrichaceae

On gravelly soil behind the dining hall at Eagle Hill campus, Steuben, ME

Description This is a typical member of the Polytrichaceae, with wide-spreading leaves on unbranched stems to 3 cm tall, but usually smaller. The upper surface of the toothed, ≈5-mm-long leaves, is flat and completely covered with lamellae, giving the plants a succulent "agave" look. Note the old antheridial cup on the right, and the densely tomentose calyptrae on the outer two capsules below. As the capsules age, they become very constricted below the mouth and they remain round in cross section, not ridged.

Similar Species Much like *P. urnigerum*, but this species is unbranched, with stems more scattered, versus occasionally branched and often in dense clumps for *P. urnigerum*. *P. dentatum* is occasionally but not exclusively alpine, whereas *P. urnigerum* is more exclusively subalpine to alpine.

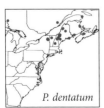

P. dentatum

Range and Habitat On gravelly soil in open disturbed sites from arctic-alpine to low elevations. More common in the northern part of our range, or in higher terrain to the south.

Name *Pogon* (G) = beard, referring to the hairy calyptra typical of the genus; *dentatum* (L) = tooth, for the coarsely-toothed leaves.

Polytrichaceae

Above, the "green earth" look of this moss. Note the tomentose capsule covers (calyptrae) and the minimal leaf growth on the much magnified photo to the right; middle right, mature capsules; bottom right, male plants. All photos from Arrowsic, ME.

Description There's really not much to *P. pensilvanicum* gametophytes. They generally grow as well-spaced plants <1 cm high and would scarcely be noticeable at all were it not for the capsules. The leaves are small, sharply pointed, toothed, and have few (10–15) lamellae. The distinctive aspect of these plants is that they rise from an extensive protonematal mat that looks like velvety green soil. This species is all about the capsules, without which the plants look like green soil. Dry capsules are round in cross section, not ridged. Note the very hairy calyptra. As with the rest of this family, the capsules begin to develop in the fall and ripen in spring.

Similar Species Potentially confused south of NY with [*P. brachyphyllum*], which is similar but has entire leaves, more lamellae, and a more crowded growth form.

4 mm

Range and Habitat Common throughout our range on disturbed soil and waste places in sun with [*P. brachyphyllum*] common in the southeast coastal plain (see map next page) and *P. pensilvanicum* more common elsewhere.

Name The *pensilvanicum* species name refers to Lancaster, PA, where the type specimen was collected

P. pensilvanicum

by Henry Muhlenberg (1753–1815), a pioneer botanist and Lutheran minister in Pennsylvania. *brachy* (G) = short, and *phyllum* (L) = leaf, for the small inconspicuous leaves. Common name: green earth.

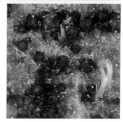

Polytrichaceae

Dense alpine zone population, Traveler Mountain, Baxter State Park, ME. See also the frontispiece photo from the same location.

Description Forming cushions to 6 cm high with occasionally branched, wiry stems, and stiff leaves 5–6 mm long × 0.5–0.7 mm wide, clustered in rosettes at the stem tips. The slightly incurved, coarsely toothed leaves have many (25–50) rows of lamellae, giving the plants a fleshy "agave" look. When dry, the leaves fold inward and upward, forming tight terminal clumps 2–3 mm across. The glaucous-green to red-brown color shown above is typical.

Similar Species *P. dentatum* is smaller, grows as more scattered plants, is usually unbranched, and is less of an alpine to subalpine specialist, though it can certainly be found at all elevations.

Range and Habitat On noncalcareous rock or gravelly soil over rock in exposed places. Generally considered an arctic-alpine species. Grout (1903), Crum and Anderson (1981), and Ireland (1982) all considered this species rare in northern New England. It's included here because it's well represented in New England herbaria. Nancy Slack (Slack and Bell, 2013) indicates that it's common along subalpine trails in New England and New York.

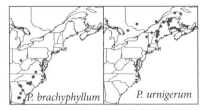

Name *urna* (L) = vessel and *gerous* (L) = bearing, indicating a capsule-producing moss, which this is only infrequently. Common name: blue-green *pogonatum*.

Polytrichaceae

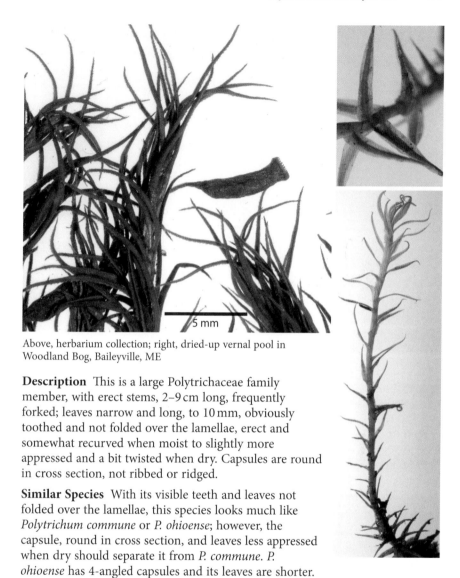

Above, herbarium collection; right, dried-up vernal pool in Woodland Bog, Baileyville, ME

Description This is a large Polytrichaceae family member, with erect stems, 2–9 cm long, frequently forked; leaves narrow and long, to 10 mm, obviously toothed and not folded over the lamellae, erect and somewhat recurved when moist to slightly more appressed and a bit twisted when dry. Capsules are round in cross section, not ribbed or ridged.

Similar Species With its visible teeth and leaves not folded over the lamellae, this species looks much like *Polytrichum commune* or *P. ohioense*; however, the capsule, round in cross section, and leaves less appressed when dry should separate it from *P. commune*. *P. ohioense* has 4-angled capsules and its leaves are shorter.

Range and Habitat Throughout the mountains in our range, south as far as North Carolina at higher elevations. On soil and soil over rock in conifer forests. Many, but certainly not all, collections are from alpine regions. Photos to the right are from low-elevation habitat.

P. alpinum

Name *poly* (G) = many, *trich* (G) = hairs, referring to the hairy calyptra of this family; *astrum* (L) = star, for the close relationship to the genera in Polytrichaceae and the spreading leaves, with *alpinum* for its preferred habitat.

Polytrichaceae

P. commune with mature capsules, hairy calyptrae, and splash cups, at Keene, NH, near airport

Description This trio of similar-looking plants represents the largest of our hair-caps at ≈3–8 cm tall, with the upright unbranched growth form and wide-spreading leaves typical of most of the family. The hairy calyptrae, ridged mature capsules, and antheridial splash cups, shown in detail below, are also typical of much of the family. When dry, the leaves of this group fold upward against the stem with little twisting or contorting (see also *Polytrichum juniperinum*). The leaves are toothed, and lamellae cover most of the upper surface of the leaves, making them fairly rigid and opaque.

Similar Species *Polytrichastrum ohioense* and *P. pallidesetum* have similar capsules, longer and narrower than *Polytrichum commune* capsules (see next page), and they are more likely to be found in shaded forests, with *P. ohioense* typically in deciduous forests, replaced by *P. pallidesetum* in more northern conifer forests. According to Crum and Anderson (1981), *P. pallidesetum* is appropriately named and can be distinguished by a seta lighter in color than the red-brown of the other two. See Polytrichaceae page for help separating these three similar species.

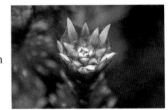

Habitat and Substrate Throughout our range on soil in open areas for *P. commune* and shaded sites for *P. ohioense* and *P. pallidesetum*, with a wide range of moisture tolerance. See maps next page.

Polytrichum splash cup

Polytrichaceae

Continued on next page

Name *poly* (G) = many, *trich* (G) = hairs, referring to the hairy calyptra of this genus; *commune* (L) = common, *ohioense* refers to the state in which that species was originally collected, and *pallidesetum* refers to the lighter-colored seta of that species. Common name for *P. commune*: common hair-cap.

Comparison Page: *P. commune* and Associates

2 mm

Left pair, *Polytrichum commune*; right pair, *Polytrichastrum pallidisetum*

P. commune   P. ohioense   P. pallidisetum

P. commune (duller, more yellow-green, toothed leaves) and *P. juniperinum* (shinier, more silver-green leaves with lamina wrapped over the lamellae). The photosynthetic cells of the lamina of the duller *P. commune* leaves are fully exposed to the light, while *P. juniperinum* leaves cover the lamellae, exposing the more reflective undersurface of the leaf lamina. Arrowsic, ME.

Juniper hair-cap with immature capsules (November), Arrowsic, ME

Description Highly variable in size, as small as 1 cm high in depauperate rock crevices, sharing space with and looking much like *P. piliferum*, or as tall as 6 cm or more when consorting with *P. commune* (see *P. commune*). This and two other hair-caps, *P. piliferum* and *P. strictum*, have folded-over leaves completely covering the lamellae. That and the short reddish leaf tip easily identify this species. The photo at left shows how dry leaves in this genus fold up against the stalk, compared with leaves in the *Atrichum* genus that contort. Immature sporophytes are produced by the Polytrichaceae in the fall and mature in June or July of the following year.

Similar Species The only other hair-caps with folded-over leaf laminae are *P. piliferum*, with much longer white leaf tips that are real awns, not simply a leaf tip extension, and *P. strictum*, which grows in bogs with *Sphagnum* spp.

Range and Habitat Throughout our range on dry gravelly soil, and soil over rock, in partial to full sun.

Name *juniperinum* for the spiky juniper-like leaves and its silver-green color. Common name: juniper hair-cap.

P. juniperum

P. piliferum in a typical site with map lichen and rock tripe. Close-up photos below show dry and hydrated plants. Lens cap is 50 mm across. Cadillac Mountain, Bar Harbor, ME.

Description The smallest of our hair-caps at 0.5–4 cm tall (usually at the small end of the range), *P. piliferum* is easily identified by the leaf edges folding over the lamellae and the long white awns rising abruptly from the leaf tips. The white hair points that give this species a hoary look are thought to protect the plant from the harsh solar rays and are a very common adaptation found in a wide variety of rock mosses that grow in full sun. Capsules are 1–2 mm long and 4-angled.

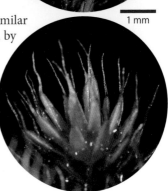

1 mm

Similar Species *P. juniperinum* can look very similar in the same sunny locations but is distinguished by its reddish leaf tips that begin gradually, versus the abrupt, long white awns of this species.

Range and Habitat Throughout our range on dry gravelly soil and soil over rock in partial to full sun.

P. piliferum

Name *pilifer* (L) = hairy, referring to the white awns.

Polytrichaceae

On a hummock with *Sphagnum fuscum,* Wonderland Bog, Bar Harbor, ME

Description A slender moss, 3–15 cm high, easily identified by habitat. Its leaves, entire and folded over the lamellae, frequently have reddish tips and are quite straight and erect even when hydrated, providing its characteristic slender look. The stem, at least the lower portion, is covered with whitish tomentum.

Range and Habitat Throughout our range on the tops of hummocks in *Sphagnum*-dominated bogs. Sometimes described as an alpine zone specialist, it actually grows in *Sphagnum* bogs at any elevation.

Similar Species Formerly named *P. juniperinum* var. *affine,* it is closely related to *P. juniperinum.* They share folded-over, entire leaves, and the frequently red leaf tips you see in the close-up photo. *P. strictum* has less divergent and more upward-pointing leaves, and stem tomentum is usually absent in *P. juniperinum.* The two species can be separated by habitat.

P. strictum

Name *strictus* (L) = drawn close together, referring to the upright growth form of the leaves.

Polytrichaceae

On gravelly soil in a calcareous sandpit, Augusta, ME; top left, wet; top right, dry; right, capsules

Description A very small moss with stems <1 cm high. The acute to acuminate leaves 1.5–2.5 mm long, keeled, with recurved margins, are erect-spreading when moist to slightly twisted and appressed when dry, with a strong, slightly excurrent costa. The papillae (see below right) on the cell surface give these leaves a dull, nonreflective look. The red stems are hidden by crowded leaves. Dioicous. The setae are red-brown, and the distinctive capsules have long, narrow, twice-twisted peristome teeth (16 teeth split into 32 segments).

Similar Species [*B. convoluta*] is similar, with yellow setae, leaf margins recurved only near the base, a subpercurrent to percurrent costa, leaves 1.2–1.5 mm long, and several peristome tooth twists. *Tortella tortuosa* is much larger with longer and more twisted teeth. See also *T. humilis*.

As seen through a compound microscope—an interesting look at the multipapillose leaf cells

B. unguiculata

Range and Habitat Throughout our range on calcareous, disturbed, gravelly soil.

Name *barba* (L) = beard, for the shape of the long, twisted peristome teeth; *unguiculus* (L) = a small nail or claw, presumably also for the twisted peristome teeth.

Above, wet (large photo) and dry (inset) (photos by Bob Klips, Marion Cemetery, Marion, OH); below, Bry. Eur.

Description Small yellow-brown plants to 1 cm tall in dense tufts with short red stems; lower stems with reddish rhizoids; leaves entire, wide spreading, obtuse to somewhat pointed with a mucronate tip formed from an excurrent costa, 2–4 mm long, wavy edged, sometimes with horizontal tears, curled and crisped when dry. As with most of this family, the leaf surface is dull and nonreflective because of papillae (bumps) on the leaf cell surfaces; monoicous, and a frequent producer of capsules ≈3 mm long with 32 long, twisted, filamentous peristome teeth; setae yellow.

Similar Species *Tortella tortuosa* is larger, has longer, more tapered leaves, and is less likely to produce capsules. [*Tortella fragilis*], also small, is less common; it has leaves frequently broken (fragile) and curled but not crisped when dry. *Barbula unguiculata* has leaves more tapered, and red setae.

T. humilis

Range and Habitat Scattered throughout our range, but more common to the south and west, on tree bases and rock and soil over rock, often calcareous.

Name *tortella* (L) = twisted, for the long and twisted capsule teeth; *humilis* (L) = small, or on the ground, for the size and low-growing form.

Near Mont Albert, Chic-Choc Mountains, Québec, Canada

Description Plants to 6 cm tall in dense tufts; stems covered with dense red-brown tomentum as shown in the lower right inset above; leaves closely packed, long and narrow, 5–6 mm long, costate, wavy edged, curled and crisped when dry, dull looking due to papillae on the leaf cell surfaces (see photo, p. 185); dioicous, capsules not common, setae reddish to pale; capsule ≈3 mm long with 32 long, twisted, filamentous peristome teeth, as shown.

Similar Species [*Tortella fragilis*], is less common and smaller, with leaves erect to incurved and curled but not crisped when dry. The *fragilis* epithet refers to commonly broken leaf tips. *Tortella humilis*, another small species, has leaves less long, narrow, and finely pointed.

Bry. Eur.

T. tortuosa

Range and Habitat Throughout eastern North America south to South Carolina and Tennessee, on calcareous soil or rock.

Name *tortus* (L) = twisted or winding, for the crisped leaves, or the capsule teeth.

On calcareous soil, along the margin of an unused dirt road through a hayfield, Unity, ME

Description Yellow-green, short, small, scattered or clustered plants to 0.5 cm tall with leaves 1.5–2.5 mm long, ovate, and with an excurrent costa creating an extended tip. Unusual for this family, the leaf cells are not papillose and should not appear dull and unreflective. Monoicous and a prolific producer of these small capsules (typically ≤1mm) that are typically produced in fall.

Similar Species *Physcomitrium pyriforme* has leaves and capsules in much the same form, a yellow seta, and entire leaves, but is much larger, with stems >0.5 cm tall, and capsules 1–2 mm long. Also, it has no soil preference, versus calcareous soil for this species, and its costa is percurrent versus excurrent for this species, though that is probably hard to see with a hand lens. *P. pyriforme* produces capsules in spring versus fall for this species

12×

Range and Habitat From Québec and the Canadian Maritimes south to

T. truncata

Maryland and Michigan on bare, disturbed, calcareous soil in the open, and at the edges of fields among grasses or on roadsides.

Name *tortula* (L) = a small twist, for the twisted peristome teeth common in closely related species, but not present in this species; *truncata* (L) = cut off, for truncate, for the capsule shape.

Note the small size and tightly contorted leaves. Capsules <1 mm long. Above, photo by Hermann Schachner; below, Bry. Eur., in Grout 1903.

Description Small, inconspicuous, yellow-green plants in tufts or mats, with red, unbranched or fork-branched stems to 5 mm high; leaves entire, acute, long-pointed, 1.5–2 mm long, erect-spreading, incurved to subtubulose, crisped and contorted when dry, dull looking due to papillae on the leaf cell surfaces (see image on *Barbula unguiculata* page), costate. Autoicous (a form of monoicous), and a reliable producer of capsules.

Similar Species The small plant size, yellow-green color, and leaves costate, entire, and very crisped when dry, on disturbed, frequently calcareous soil should be a good start on confirming this species.

Range and Habitat Throughout our range, and from Newfoundland to Florida on exposed bare soil in disturbed areas or on soil over rock. Uncommon and probably a calciphile (as are most Pottiaceae) in the north while very common and more of a generalist in the southern part of our range.

W. controversa

Name *Weissia* is for Friedrich Wilhelm Weiss (1744–1826), who wrote *Illustratio Systematis Sexualis Linnaeani*; *controversa* is for nomenclature difficulties dating to Johannes Hedwig's *Species Muscorum* (1803).

Pottiaceae   Ca in the north

All photos from under a tipped-up tree base in Raquette Lake, NY

Description The charismatic goblin's gold is uncommon and unusual. This is the only species in the order Schistostegales, the family Schistostegaceae, and the genus *Schistostega*. It produces male and female gametophytes on separate plants rising from a common protonema, making its sexual condition (dioicous versus monoicous) unclear. The protonema has a reflective inner layer maximizing the low levels of light it receives and providing the delightful appearance of luminescence. The sterile leaves (next page) are gray-green, flat, and pinnate. I haven't seen capsules, but they are short and round, and I'm not sure how commonly they are produced.

This uncommon species is included here for the simple reason that it's great fun to seek and more fun to find. Good luck.

Similar Species *Fissidens* spp. are superficially similar to the flat sterile leaves shown on the next page, but this species has none of the extra leaf flaps found on *Fissidens* spp., and *Fissidens* spp. lack a reflective protonema.

Range and Habitat Uncommon, but might be encountered in the northern part of our range from upstate NY across Vermont and NH to Maine, and throughout boreal Canada under tree tip-ups, in cave mouths, under rock overhangs, in cellar holes, sheds, or any dark place, on mineral soil. Keep looking.

S. pennata

Name *chostos* (G) = cleft, *schist* (L) = slate, a splitting rock; *stege* (G) = shelter or roof, indicating the cave mouth or rock overhang habitat, *penna* (L) = feather, for the pinnate shape of the gametophyte leaf. Some authors [Allen (2005) and Crum and Anderson (1981)] suggested that the genus name refers to a strangely splitting operculum, which this species lacks. Common name: goblin's gold.

Schistostegaceae   Continued on next page

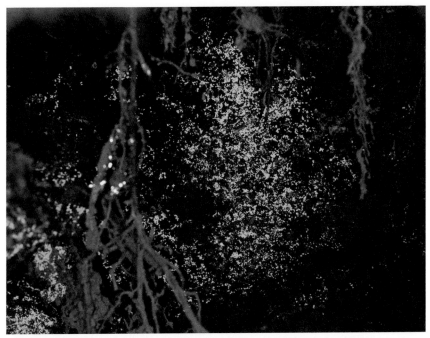

Above, the reflective protonema appears to glow; below, sterile leaves on gravelly soil under the tip-up

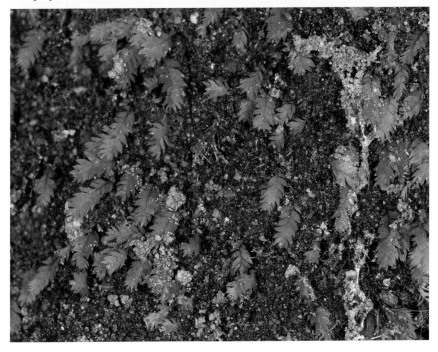

Schistostegaceae

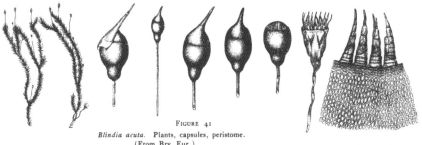

FIGURE 41

Blindia acuta. Plants, capsules, peristome.
(From Bry. Eur.)

5 mm

In drawings at top, plants on left are life-size, capsules are ≈1 mm long and drawn at 10× (Bry. Eur., in Grout 1903); above, photo by Hermann Schachner, courtesy of Wikimedia Commons

Description A very small moss with stems to 4 cm high, forming shiny black to dark green cushions on wet rocks. Stems red and repeatedly forked. Leaves narrow and *Dicranum*-like, ≈2 mm long, slender, subulate to subtubulose, erect to erect-spreading and somewhat incurved when dry. The excurrent costa fills the upper leaf lamina. The pyriform capsules, much illustrated above, are unfortunately rare. Dioicous.

Similar Species *Dicranum fulvum,* the rock *Dicranum,* is more falcate-secund and doesn't share the wet rock habitat.

Range and Habitat Throughout the Northeast, reported south to PA, with reports from higher terrain in NC and WV. On acidic rocks in streams, seeps, wet ledges, or around waterfalls. Nancy Slack (pers. comm.) says it's common in streams and on lakeshores in the Adirondacks. Grout (1903) said it's rare, Allen (2005) reported it from eight Maine counties, and Ireland (1982) said it's common in suitable sites in the Canadian Maritimes. This seems to be a northern and arctic-alpine species that is rare south of New England.

B. acuta

Name *Blindia* is for Pastor J. J. Blind of Münster, Germany, and *acuta* is for the long, sharply pointed leaves.

Seligeriaceae

2 mm

S. pensylvanicum on moose dung in Big Heath, Mount Desert Island, ME

Description This genus of light green mosses grows on dung or carcass remains, usually in bogs. They are rare, and therefore probably shouldn't be in this book, but they're fun to look for if you find yourself in a boggy environment where a moose or deer could have left droppings, or might have died. Of the four species that might be found in our region, three are restricted to the extreme northern part of the United States and farther north through the Canadian Arctic and Alaskan North Slope. Only *S. pensylvanicum*, shown above, can be found throughout eastern United States, occasionally on bovine dung.

Splachnum protonema on moose dung.

Similar Species *S. rubrum* and *S. luteum* have umbrella-shaped capsules and are species of the extreme north, though they have been found in the northern part of our U.S. range. *S. ampullaceum* (lower right) has a capsule somewhat like those shown above but with more swelling in the neck area. It has been recorded in northern New England and upstate NY and is probably more common than *S. pensylvanicum*, though I've never found it.

Mary V. Thayer illustration in Grout 1903.

Range and Habitat In bogs and pastures. See Similar Species for range discussion. Rare.

S. pensylvanicum

Name *Splachn* (G) = entrails or viscera, probably for the required nitrogen-rich substrate.

Splachnaceae

Having completely occupied this stump, this moss colony is now producing many spore capsules to find a new home. Lower right, a gemma cup, this moss's way of reproducing vegetatively, presumably for short-distance travel. Ravena, NY.

Description Small plants with short-stalked, occasionally forked stems either sprawling or upright, on rotting wood or stumps. Leaves appropriately pellucid, ovate to ovate-lanceolate, ≈2 mm long, frequently formed into cups containing asexual reproductive gemmae (below right). Once a rotting stump is completely covered with stems, the plants will produce copious quantities of unusual 4-toothed capsules with red-orange setae, giving the stump a distinctive orange look, as shown above.

M. V. T.

Similar Species Either the gemma cups or the unusual 4-toothed capsules make this an easy ID.

Range and Habitat Common throughout North America, south in the east to Florida. Found primarily on rotting wood, occasionally on sandstone in the south.

T. pellucida

Name *Tetra* (G) = four, and *fissus* (L) = split, for the capsule with 4 teeth; *pellucidus* (L) = transparent, or to shine through, for the translucent leaves.

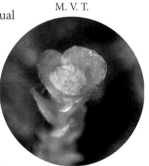

Gemma Cup

Tetraphidaceae

Pleurocarpous Mosses

Unlike the acrocarps, which are almost all costate, pleurocarps can be either costate or not. They grow in mats, with a main stem prostrate or nearly so, sometimes with erect tips or branches. Pleurocarps are usually much branched, often pinnate, rarely simple, and if so, the stems are long and intertwined. Sporophytes rise from the tip of a small, very reduced, often inconspicuous branch along the main stem well below the apex of a major branch.

Some pleurocarps have tightly packed unbranched stem tips, and they can look very much like the cushion- or tuft-forming acrocarps. Some teasing apart of densely packed pleurocarps will usually show branching from a horizontal main stem. *Anomodon* and *Entodon* spp. are good examples of tuft-forming pleurocarps that might be confusing.

Brachythecium, Hypnaceae, and Hook Mosses (former *Drepanocladus* genus) are given special treatments in this section.

SPECIES INCLUDED

The order followed here, alphabetically by family, reflects the order of the species accounts. Species in [brackets] are mentioned in keys, but not illustrated.

Amblystegiaceae (p. 221)
  Calliergon cordifolium
  Straminergon stramineum
  Calliergon giganteum

Campylium hispidulum
Campylium stellatum
Campyliadelphus chrysophyllus
Hygroamblystegium fluviatile

Hygrohypnum ochraceum
Leptodictyum riparium
[*Platylomella lescurii*]
Sanionia uncinata
Warnstorfia exannulata
Anomodontaceae (p. 233)
Anomodon attenuatus
Anomodon rostratus
Anomodon viticulosus
Haplohymenium triste
Brachytheciaceae (p. 237)
Brachythecium campestre
[*Brachythecium reflexum*]
Brachythecium rivulare
Brachythecium rutabulum
[*Brachythecium velutinum*]
Bryhnia novae-angliae
Bryoandersonia illecebra
[*Eurhynchium pulchellum*]
Pseudoscleropodium purim
Tomentypnum nitens
Climaciaceae (p. 245)
Climacium americanum
[*Climacium dendroides*]
Entodontaceae (p. 246)
[*Entodon cladorrhizans*]
Entodon seductrix
Fontinalaceae (p. 247)
Dichelyma capillaceum
Fontinalis antipyretica
Fontinalis dalecarlica
Fontinalis novae-angliae
Hedwigiaceae (p. 251)
Hedwigia ciliata
Hylocomiaceae (p. 252)
Hylocomiastrum umbratum
Hylocomium splendens
Loeskeobryum brevirostre
Pleurozium schreberi
[*Rhytidiadelphus loreus*]
[*Rhytidiadelphus squarrosus*]
Rhytidiadelphus triquetris
Rhytidium rugosum
Hypnaceae (p. 258)
Callicladium haldanianum

Calliergonella cuspidata
Herzogiella striatella
[*Herzogiella turfacea*]
Hypnum cupressiforme
Hypnum cupressiforme var.
filiforme
Hypnum curvifolium
Hypnum imponens
Hypnum lindbergii
Hypnum pallescens
Platygyrium repens
Pseudotaxiphyllum elegans
Ptilium crista-castrensis
Pylaisia polyantha
Taxiphyllum deplanatum
Leskeaceae (p. 274)
Leskea polycarpa
Leskeella nervosa
Leucodontaceae (p. 276)
Forsstroemia trichomitria
[*Leucodon brachypus*]
[*Leucodon julaceus*]
Leucodon sciuroides
Neckeraceae (p. 279)
Homalia trichomanoides
Neckera pennata
Thamnobryum alleghaniense
Plagiotheciaceae (p. 282)
Plagiothecium cavifolium
Plagiothecium denticulatum
Plagiothecium laetum
Pterigynandraceae
[*Pterigynandrum filiforme*]
Sematophyllaceae (p. 285)
Pylaisiadelpha recurvans
Theliaceae (p. 286)
[*Myurella julacea*]
Myurella sibirica
Thelia hirtella
Thuidiaceae (p. 288)
Abietinella abietina
Helodium blandowii
[*Heterocladium dimorphum*]
Thuidium delicatulum
[*Thuidium recognitum*]

DICHOTOMOUS KEY TO GENERA AND SPECIES OF PLEUROCARPOUS MOSSES

1. Leaves strongly rugose when dry, falcate-secund, margins recurved; large plants with large leaves; on dry, exposed, calcareous rocks and cliffs.............. .... *Rhytidium rugosum* (p. 257)
1. Not as above 2

2. Plants complanate, leaves appearing distichous, oblong, falcate, apices rounded; on calcareous rock ledges, bluffs, shaded faces of cliffs, and occasionally bases of trees............................ *Homalia trichomanoides* (p. 279)
2. Plants not complanate, or if so, leaves not oblong with rounded apices........3

3. Leaves very small (<1 mm) and mostly broken except at tips of branches, plants small and stringy, on hardwood tree trunks.. ....*Haplohymenium triste* (in part) (p. 236)
3. Leaves not as above 4

4. Leaves with a single costa..5
4. Leaves lacking costa or costa double (usually divided at base and not visible with hand lens).. 35

Costate Pleurocarps

5. Plants dendroid and erect 6
5. Plants not dendroid...7

6. Branches covered with paraphyllia that give them a whitish or brownish cobwebby appearance; leaves shiny; in swampy habitats on wet soil and humus in swamps, along streams, beside pools in woods, and on shady, damp rocks ....*Climacium americanum*/[*dendroides*] (p. 245)
6. Branches lacking paraphyllia; leaves dull; plants on wet rock beside creeks and on humus over rock, not in swampy habitats .... *Thamnobryum alleghaniense* (p. 281)

7. Apices of many branches with clusters of microphyllous branchlets; small plants on tree trunks, rotten logs, and calcareous rocks ....*Leskeella nervosa* (p. 275)
(Note: *Leucodon sciuroides* and *Platygyrium repens* also grow on tree trunks and have microphyllous branchlets but are ecostate.)
7. Apices of branches lacking microphyllous branchlets.....................................8

8. Stems pinnate to tripinnately branched, sometimes frondose; stems and branches covered with paraphyllia (white, green, yellow, or brown with a hairy, granular, or cobwebby appearance) 9
8. Stems irregularly branched or rarely pinnately branched but never frondose; paraphyllia lacking or not evident (rhizoids sometimes present that somewhat resemble paraphyllia, but they occur only on the main stem)........ 12

9. Stems 1-pinnate, with a cobwebby appearance caused by a dense covering of long paraphyllia; plants of wet habitats, especially swampy cedar woods and calcareous swamps, fens and pools in woods..
.... *Helodium blandowii* (p. 289)

9. Stems 1–3-pinnate, with a hairy or granular appearance due to short paraphyllia; often in dry habitats 10

10. Plants 1-pinnate 11

10. Plants 2–3-pinnate.. *Thuidium*

Thuidium

A. Stem leaves arched and standing out from stem when dry; on calcareous soil, humus, boulders, and sometimes bases of trees in woods....................
.... [*T. recognitum*] (p. 290)

A. Stem leaves appressed or slightly spreading from stem when dry; on acidic substrata in wet habitats such as humus, soil, boulders, rotten logs, and stumps in woods.. *T. delicatulum* (p. 290)

11. Large plants; leaves lacking twisted tips, with costa 3/4 of leaf length; on dry, exposed, calcareous rocks and cliff shelves, on humus on slopes, and on rotten stumps... *Abietinella abietina* (p. 288)

11. Large plants; leaves with twisted tips, with costa extending 1/2 to 2/3 leaf length; often in mats in damp grassy areas ....
.... *Pseudoscleropodium purim* (in part) (p. 243)

12. Margins of stem and branch leaves ciliate; on bases of deciduous trees, particularly white oak .... *Thelia hirtella* (p. 287)

12. Margin of stem and branch leaves not ciliate................................... 13

13. Stems much branched, covered with a dense mat of brown rhizoids that are lacking on branches; in calcareous swamps and fens, very large.....................
.... *Tomentypnum nitens* (p. 244)

13. Stems irregularly branched or if subpinnately branched, without dense rhizoids; plants of various habitats, small to large.. 14

14. Leaves squarrose; on calcareous or noncalcareous rocks and soil, rotten wood and bases of trees (squarrose species without costa begin at couplet 59) .... *Campyliadelphus chrysophyllus* (p. 225)

14. Leaves not squarrose 15

15. Stem leaves with an obtuse apex 16

15. Stem leaves with an acute apex.. 19

16. Stem leaves large, >1 mm long, cupped, with twisted tips; plants large............
.... *Pseudoscleropodium purim* (in part) (p. 243)

16. Stem leaves small or large, without twisted tips 17

17. Stem leaves very small, <1 mm long, with broken tips.....................................
.... *Haplohymenium triste* (in part) (p. 236)

17. Stem leaves >1 mm long, without broken tips, often broad and entire........ 18

18. Plants of wet habitats (bogs, swamps, fens) and with leaves usually spreading except at tip ....
.......*Calliergon/Straminergon* (see *Calliergonella* [p. 259] for ecostate relative)

Calliergon/Straminergon

A. Stems often pinnate with numerous branches; stem leaves broad, sometimes nearly as broad as long; in bogs, fens, swamps, shallow pools, ponds, and near springs ....*C. giganteum* (p. 222)

A. Stems irregularly branched with few branches; stem leaves narrow, clearly longer than broad; in bogs, swamps, drainage ditches, and wet depressions.. B

B. Stem leaves with brown rhizoids at the tips of at least some leaves (particularly see older, lower leaves), probably a calciphile.........................
....*S. stramineum* (p. 221)

B. As above without brown rhizoids, and not requiring high pH ....
....*C. cordifolium* (p. 221)

18. Plants of dry habitats, always on limestone..... *Anomodon viticulosus* (p. 235)

19. Leaves strongly falcate-secund, especially at stem and branch tips.............. 20
19. Leaves not falcate-secund.. 22

20. Leaves short, seldom >1 mm long...
............[*Brachythecium velutinum*] (see also *Brachythecium* key on page 237)
20. Leaves longer, mostly >1 mm long.. 21

21. Leaves long-acuminate, smooth; perichaetial leaves long and sheathing, nearly reaching capsule; on bases of trees, shrubs, and rocks beside streams in periodically flooded places at margins of ponds and lakes..........................
....*Dichelyma capillaceum* (p. 247)
21. Leaves long-acuminate; stem leaves sometimes plicate; perichaetial leaves short and not sheathing the setae; often in swamps, fens, and bogs or sometimes in woodlands on rock, wood, or humus...
.... Hook Moss group (see Hook Moss key, p. 229, including the genera *Sanionia* and *Warnstorfia*)

22. Plants on rock, usually in streams... 23
22. Plants on various substrates and if on rock, not in streams........................ 27

23. Leaves narrow, acuminate, <1 mm wide 24
23. Leaves broad, often >1 mm wide.. 26

24. Stems often with a wiry appearance, the basal part of the stem with leaves lacking and often only the remnants of costae attached 25
24. Stems without a wiry appearance, the entire leaves intact................................
.... *Leptodictyum riparium* (p. 238) (see also couplet 27)

25. Leaves with a differentiated marginal border of a different color and thickness than the rest of the leaf; on rocks and boulders in waterfalls, creeks, and rivers ....[*Platylomella lescurii*]

25. Leaves without differentiated marginal border; on calcareous and noncalcareous boulders, rock ledges, and bluffs, or rarely woody debris in creeks and rivers .... *Hygroamblystegium fluviatile* (p. 226)

26. Leaves decurrent .... *Brachythecium rivulare* (p. 239)
26. Leaves nondecurrent; on soil over acidic rocks in and beside creeks, streams, and waterfalls, sometimes beside lakes..
.... *Hygrohypnum ochraceum* (p. 227)

27. Leaves long in relation to width, 7–15× as long as wide; plants usually in aquatic habitats such as swamps, creeks, and rivers on rocks and boulders (sometimes calcareous), also on fallen branches and woody debris in swamps and stagnant pools............................. *Leptodictyum riparium* (p. 228)
27. Leaves short in relation to width, mostly <7× times as long as wide; plants mostly of dry or mesic habitats 28

28. Leaves dull when dry; costa bulging on dorsal leaf surface and of a different color from the lamina 29
28. Leaves glossy when dry; costa similar in color to lamina.............................. 31

29. Leaves acute to nearly obtuse; on bases of trees or decaying logs, often in periodically flooded places.. *Leskea polycarpa* (p. 274)
29. Leaves ending in an apiculus or a long, hyaline hair point........................... 30

30. Leaves (at least many) with a long, hyaline, often toothed apex......................
.... *Racomitrium* (see Grimmiaceae in Acrocarpous section)
30. Leaves apiculate or with a long smooth hair point ....*Anomodon*

Anomodon (see also *Haplohymenium triste*, couplets 3 and 17)

A. Leaves with a long, filiform, hyaline acumen; leaf margins revolute; on rocks and in cliff crevices that are frequently calcareous, on bases of trees, and sometimes on soil or humus ....*A. rostratus* (p. 234)
A. Leaves apiculate, acute, or obtuse; leaf margins plane.............................. B
B. Plants with attenuate branches (especially noticeable when dry); branch leaves gradually narrowed to an acute apex, scarcely contorted when dry; occurring on tree trunks and bases, rotten stumps, calcareous and noncalcareous rock, and cliff shelves........................ *A. attenuatus* (p. 233)
B. Plants without attenuate branches; branch leaves obtuse or apiculate, strongly contorted when dry; on calcareous rocks and cliff ledges, rarely on tree trunks..*A. viticulosus* (p. 235)

31. Branches julaceous and leaves concave .... *Bryoandersonia illecebra* (p. 242)
31. Branches not julaceous, with leaves flat 32

32. Branch leaves broad at apex, acute to narrowly obtuse, strongly serrate, not twisted; stem leaves more acuminate and sometimes with twisted tips; on soil, rotten stumps and logs, bases of trees, rock outcrops, and humus over rock .... [*Eurhynchium pulchellum*]
32. Branch leaves narrow at apex, acute to acuminate, sometimes twisted....... 33

33. Stem leaves smooth, nearly as broad as long, apex often abruptly narrowed, acute and twisted; on soil, rocks, humus, and rotten logs in wet shady places, especially along creeks ....*Bryhnia novae-angliae* (p. 241)
33. Stem leaves often plicate, usually longer than broad, apex usually gradually narrowed, acuminate, straight, or sometimes twisted 34

34. On tree trunks..*Forsstroemia trichomitria* (p. 276)
34. On tree bases, soil, rock, or soil over rock.................... *Brachythecium* (p. 237)

Ecostate Pleurocarps

35. Plants aquatic, in flowing water of streams; stems long, often 10 cm or more in length ....*Fontinalis*

Fontinalis

A. Leaves keeled; plants often yellowish to brownish green; on boulders and twigs in creeks and ponds, usually in moving water ....
.... *F. antipyretica* (p. 248)
A. Leaves concave; plants usually green to brownish .... B

B. Leaves broad, 2–3× as long as wide, the margins plane when dry; on rocks in flowing or still water in creeks and rivers or at margins of lakes
....*F. novae-angliae* (p. 250)
B. Leaves narrow, 3–5× as long as wide, the margins sometimes reflexed when dry; on rocks, branches and logs in running water of creeks and rivers.. *F. dalecarlica* (p. 249)

35. Plants not aquatic 36

36. Leaves with whitish tips, those surrounding immersed capsules with cilia on margins; capsules without teeth; on dry, exposed, usually acidic boulders and cliffs...*Hedwigia ciliata* (p. 251)
36. Leaves lacking whitish tips and cilia; peristomate; on various substrates ... 37

37. Stems and branches with clusters of microphyllous branchlets in leaf axils....
.... 38
37. Stems and branches lacking clusters of microphyllous branchlets in leaf axils 40

38. Leaves distant, wide-spreading, complanate; stems and branches often visible between leaves; on soil and humus on banks and cliffs in moist woodlands..*Pseudotaxiphyllum elegans* (p. 270)
38. Leaves close, erect; stems and branches not visible between leaves; on trees..
.... 39

39. Sometimes with sporophytes; leaves smooth; branches straight when dry; olive green; on tree trunks, rotten logs and stumps...
.... *Platygyrium repens* (p. 269)
39. Never with sporophytes; leaves plicate; branches curved upward when dry; dark green; on deciduous tree trunks or rotten logs...
.... *Leucodon sciuroides* (p. 277)

40. Stems and branches covered with long, branched, white, yellow, or green paraphyllia, giving them a cobwebby appearance; stems often frondose, 1–3 pinnate ....*Hylocomium*

Hylocomium/Hylocomiastrum

A. Stems regularly branched, 2–3 pinnate; stem leaves usually with a long, slender, undulate acumen; on humus, rotten logs, soil, and in swamps and forests..*Hylocomium splendens* (p. 253)

A. Stems irregularly branched, 1–2 pinnate; stem leaves broadly acuminate and not undulate; on rotten wood and humus over rocks in forests..........
....*Hylocomiastrum umbratum* (p. 252)

40. Stems and branches lacking paraphyllia 41

41. Plants julaceous, at least at stem tip.. 42
41. Plants not julaceous.. 49

42. Plants pinnately branched 43
42. Plants irregularly branched 45

43. Plants small; stem leaves acuminate, seldom up to 1 mm long; on soil and humus banks, bases of trees and boulders in moist woods ....
....[*Heterocladium dimorphum*]
43. Plants robust; stem leaves apiculate, often ≥1 mm long.................................. 44

44. Stems and branches orange or red; on soil and humus in woods, occasionally in bogs..*Pleurozium schreberi* (p. 255)
44. Stems and branches yellow or green; a calciphile occurring in swamps, fens, and alkaline bogs .... *Calliergonella cuspidata* (p. 259)

45. Plants large, stem leaves often ≥1 mm long.. 46
45. Plants small, stem leaves <1 mm long 48

46. Primarily on deciduous tree trunks, occasionally on rotten logs; branches curved upward when dry; plants glossy, leaves smooth, capsules exserted .....
....[*Leucodon julaceus*] (p. 277)
46. Plants on soil or humus over rock, sometimes on bases of trees and rotten wood, with straight, or at least not upcurved, branches 47

47. Leaves often short to long-acuminate, contorted when dry; stems often visible between leaves; on soil or humus over rock, sometimes on bases of trees and rotten wood ....*Plagiothecium* (see also couplet 53).

Plagiothecium

A. Plants julaceous to complanate; leaves concave, symmetrical, the short tips often recurved; capsules erect, often striate when dry; on usually calcareous soil over cliff ledges, on stumps, rotten wood, bases of trees, clay banks, and humus in woods.................................*P. cavifolium* (p. 282)

A. Plants complanate; leaves flat, asymmetrical, the tips not recurved; capsules erect or inclined, smooth or striolate when dry ....B

B. Leaf margins recurved; capsules inclined, striolate when dry; in wet woods on rotten logs, soil, humus, and rarely on rocks................................
....*P. denticulatum* (p. 283)

B. Leaf margins plane; capsules erect, rarely inclined, smooth; common on rotten logs, stumps, bases of trees, humus and soil on steep banks and over boulders and cliffs in woods.......................................*P. laetum* (p. 283)

47. Leaves acute to apiculate, scarcely contorted when dry; stems not visible between leaves; on bases of trees and rotten wood in deciduous woods, rarely on rock.. *Entodon seductrix* (p. 246)

48. Leaves about as broad as long, often ending in a short, hairlike apiculus; margins sometimes spinulose; on calcareous rock............................... *Myurella*

Myurella

A. Plants julaceous; leaves close, imbricate, apiculate or sometimes obtuse; margins entire or nearly so................ [*M. julacea*] (see *M. sibirica*, p. 286)

A. Plants not or rarely somewhat julaceous; leaves distant, spreading, acuminate; margins often spinulose; in crevices and on soil over ledges of calcareous cliffs.. *M. sibirica* (p. 286)

48. Leaves longer than broad, acute to obtuse; margins entire; on acidic boulders and cliffs in woods or occasionally on logs and tree trunks..............
.... [*Pterigynandrum filiforme*]

49. Plants complanate... 50
49. Plants not complanate... 55

50. Leaves strongly undulate; setae short, capsules immersed; occurring on tree trunks..*Neckera pennata* (p. 280)

50. Leaves not undulate; setae long, capsules exserted; occurring on various substrates... 51

51. Leaf apices broad and rounded..................... *Homalia trichomanoides* (p. 279)
51. Leaf apices narrow and acute.. 52

52. Stems and branches orange or red; on rotten logs, bases of trees, humus, and rock...*Pylaisiadelpha recurvans* (p. 285)

52. Stems and branches yellow or green... 53

53. Leaves asymmetrical.. *Plagiothecium* (see couplet 47)
53. Leaves symmetrical.. 54

54. Leaves acuminate; usually on wet soil over calcareous rock bluffs, sometimes on bases of trees or rotten wood..
.... *Taxiphyllum deplanatum* (p. 273)

54. Leaves acute or rarely acuminate; on rotten wood and bases of trees, occasionally occurring on soil or rock, usually in deciduous woods..............
.... [*Entodon cladorrhizans*] (see *Entodon seductrix*, p. 246)

55. Leaves plicate to striolate.. 56
55. Leaves smooth, not plicate or striolate... 59

56. Primarily on deciduous tree trunks, occasionally on rotten logs; branches curved upward when dry; leaves plicate, capsules at least partially inserted among perichaetial leaves.................................. [*Leucodon brachypus*] (p. 277)
56. Not on tree trunks or rotten wood 57

57. Plants irregularly branched, leaves striolate when dry; a calciphile in fens and meadows, at margins of lakes, rarely on wet rocks ....*Campylium stellatum* (p. 224)
57. Plants pinnately branched, leaves plicate 58

58. Plants plumose; stems and branches yellow or green; leaves falcate-secund, long-acuminate; on soil or humus, primarily in coniferous forests................... .... *Ptilium crista-castrensis* (p. 271)
58. Plants sparsely pinnate; stems and branches orange or red; leaves erect or some squarrose...*Rhytidiadelphus* and *Loeskeobryum*

Rhytidiadelphus and *Loeskeobryum*

A. Stem leaves strongly squarrose when wet.. B
A. Stem leaves not strongly squarrose when wet... C

B. Stem and branch leaves smooth, strongly squarrose to squarrose-recurved; on humus, soil, rotten logs, and wet boulders in wet meadows, or sometimes on sandy soil beside rivers and lakes (rare) ....[*R. squarrosus*] (see *R. triquetris*, p. 256)
B. Stem and branch leaves mildly plicate, stem leaves strongly squarrose to squarrose-recurved when wet, branch leaves divergent; on soil, soil over rock, rotten wood, or tree bases................................*L. brevirostre* (p. 254)

C. Leaves rugose near apex, noticeably crowded near stem apices; costae strong, extending to middle of leaves and above; on humus, soil, and rotten logs in dry to moist woods, sometimes in swamps and on calcareous boulders and cliff ledges (common)..........*R. triquetris* (p. 256)
C. Leaves neither rugose nor noticeably crowded at stem apices; costae lacking or weak and ending below middle of leaves; on logs, humus, and rocks in coniferous woods (primarily a northwestern species, occasionally found in the Canadian Maritimes) ....[*R. loreus*]

59. Leaves squarrose 60
59. Leaves not squarrose 63

60. Plants subpinnately branched, stems and branches orange or red .... *Rhytidiadelphus squarrosus* (see couplet 58, and p. 256)
60. Plants irregularly branched, stems and branches yellow or green 61

61. Leaves large, often ≥2 mm long, strongly twisted when dry; a calciphile of wet habitats ....*Campylium stellatum* (see couplet 57, and p. 224)
61. Leaves small, mostly <2 mm long, straight or somewhat contorted when dry; plants of mesic to dry habitats 62

62. Leaves mostly <1 mm long, ovate with acuminate tips; capsules smooth; on soil, rocks, bases of trees, and rotten logs...
.... *Campylium hispidulum* (p. 223)
62. Leaves usually ≥1 mm long, oblong-lanceolate to ovate; capsules striate.........
.... *Herzogiella*

Herzogiella

A. Leaves close, squarrose to squarrose-recurved; on humus, acidic rocks, soil over rocks, clay banks, rotten logs, and bases of trees...........................
.... *H. striatella* (p. 260)
A. Leaves distant, erect-spreading to wide-spreading; on rotten logs, stumps, bases of trees, humus over boulders, and soil in moist coniferous woods... [*H. turfacea*]

63. Leaves falcate-secund, the apices turned toward the substrate ....
....*Hypnum* (see *Hypnum* key, p. 261)
63. Leaves straight or somewhat curved.. 64

64. Stems and branches somewhat flattened; leaves acute to short acuminate; alar cells often colored orange or brown; capsules inclined and curved; common on rotten wood, sometimes on bases of trees, soil over rock...........
.... *Callicladium haldanianum* (p. 258)
64. Stems and branches not flattened; leaves long-acuminate; alar cells not colored; capsules straight; on deciduous tree trunks 65
See discussion of differentiated alar cells on page 17

65. Plants small, branches extending <1 cm from substrate, leaves <1 mm...........
....*Pylaisia polyantha* (p. 272)
65. Plants large and robust, branches often curving 4 or 5 cm away from substrate, leaves 1.5–2 mm long...................*Forsstroemia trichomitria* (p. 276)

QUICK LOOK AT FAMILIES BY HABITAT

The pleurocarpous families, closely related to each other, provide fewer macroscopic characteristics and less capsule differentiation than do the acrocarpous families. For this reason, the Habitat Key and Key to Families are combined to shorten the keys and, it is hoped, simplify the task.

Please refer to the family descriptions that follow this section for more detail.

Some mosses show a marked loyalty to their primary habitat and substrate, and ecological characteristics can be very helpful in their identification. Then again, some species are generalists and can show up in the strangest places. We'll focus here on the most likely place to find a particular family, with no promise of including all possibilities. Please remember to check the acrocarp keys if your sample isn't clearly pleurocarpous. Habitats are listed from wet to dry.

Aquatic

Fontinalaceae: These aquatic species have a single attachment point and may be found in still or moving water. See also *Dichelyma* in Amblystegiaceae for a seasonally inundated species.

Bogs

Ombrotrophic bogs, which have minimal groundwater flow and receive their nutrients from the sky, are the province of the peat mosses (Sphagnaceae), but some mosses in the next section (Very Wet), particularly Amblystegiaceae, may also be found here.

Very Wet, Occasionally Submerged, Mucky

Amblystegiaceae: Large, much-branched plants; leaves usually pointed; not complanate; most species costate.

A few exceptions in this large family: *Campylium* spp. are somewhat drier sited and are small to medium-sized; *Campylium hispidulum* and *C. stellatum* are ecostate; and *Hygrohypnum ochraceum* has a sometimes visible double costa.

Some other, less likely, possibilities:
Brachythecium rivulare may be found here.

Leskeaceae: Small costate mosses often found in floodplain areas, so may show some signs of seasonal inundation, though they are typically on tree trunks and occasionally tree bases.

Plagiotheciaceae: Very complanate and shiny, often on wet, seepy rock faces.

Thuidiaceae: See Forest Floor section.

Very Wet, Occasionally Submerged, Rock

Not a major pleurocarpous habitat. See Acrocarpous Mosses.

Fissidentaceae: Not a pleurocarpous family, but with at least two common species that have a pleurocarpous-appearing seta and capsule. Often found on lakeshore rocks.

Forest Floor

Mosses are more common and conspicuous in conifer forests, where they aren't covered annually by fallen leaves, and where soil pH is often acidic, a condition that makes life more difficult for their competitors—the vascular plants.

1. Plants much branched, irregularly to neatly singly pinnate; leaves shiny, ecostate, falcate-secund.................................. Hypnaceae and Sematophyllaceae
1. Randomly branched to pinnate or bipinnate; leaves ecostate or costate; not falcate-secund 2

2. Very upright and dendroid (treelike), on mucky soil .... Climaciaceae
2. Not dendroid 3

3. Leaves dull, nonreflective (papillose leaf cells); randomly branched to pinnate or bipinnate..Thuidiaceae
3. Leaves shiny, reflective (smooth leaf cells); misc. branching patterns...............4

4. Complanate; leaves wide spreading; on seepy rock or mineral soil.....................
....Plagiotheciaceae
4. Not complanate, or if slightly complanate, julaceous...5

5. Costate with costa extending from 1/2 to 3/4 of leaf length; plants large (≤5 cm long) and much branched; leaves usually wide spreading, often plicate.. Brachytheciaceae
5. Ecostate..6

6. Large (stems ≤5 cm long), much branched, often pinnate to bipinnate, shiny plants; leaves appressed to squarrose, variously wrinkled..........Hylocomiaceae
6. Medium to large (stems ≤5 cm long), randomly branched, not complanate to slightly complanate plants; leaves julaceous (tightly appressed), smooth, and concave ....Entodontaceae

Rotten Stumps and Logs or Humus

As wood decays into humic soil, almost any of the forest floor species can show up, but a few species are rotten-wood specialists.

1. Leaves costate 2
1. Leaves ecostate 3

2. Plants large (≤5 cm long) and much branched; leaves with costa extending from 1/2 to 3/4 of the leaf length; leaves wide spreading, often plicate.............
.... Brachytheciaceae
2. Plants small, with stem (including leaves) <1 mm diameter; leaves ≤1 mm long, in tangled mats on tree trunks, tree bases, or humic soil........ Leskeaceae

3. Small plants; leaves <1 mm long, squarrose; with a preference for rotting wood in rich sites...........................*Campylium hispidulum in* Amblystegiaceae
3. Larger plants with leaves >1 mm long, not squarrose.......................................4

4. Large golden-tinged plants with stems ≤8 cm long; leaves straight, not falcate-secund; branches frequently appear somewhat flattened ....
....*Callicladium haldanianum* in Hypnaceae
4. Medium-sized to large green plants with random or pinnate branching; leaves falcate-secund................. *Pylaisiadelpha, Hypnum, Ptilium* in Hypnaceae

Tree Bases (up to about 0.5 m above ground)

Any of the rotten stump and log species, tree trunk species, and some soil species can occasionally be found on tree bases, particularly the lower and more horizontal parts.

1. Leaves ecostate 2
1. Leaves costate 3

2. Plants medium-sized to large, much branched, often pinnate, shiny, ecostate .... Hypnaceae

> The *Hypnum* genus, common on tree bases, has falcate-secund leaves. *Callicladium haldanianum* is another very common tree base moss in this family—see Rotten Stumps and Logs or Humus.

2. Large plants, much branched, often pinnate to bipinnate; leaves not falcate-secund, often wrinkled, plicate, or squarrose ....Hylocomiaceae

> These are mostly forest floor mosses, often moving up tree bases from the forest floor. *Hylocomium splendens*, the quite distinctive "stair-step" moss, is common on tree bases.

3. Plants shiny; large (≤5 cm long); costa extending from 1/2 to 3/4 of the leaf length; much branched; leaves wide spreading, often plicate; capsules stocky and dark .... Brachytheciaceae

3. Plants shiny or dull, large or small; costa extending to leaf tip; branching variable; leaves usually appressed 4

4. Small plants, with stem (including leaves) <1 mm diameter; leaves ≤1 mm long, not dull; in tangled mats on tree trunks, tree bases, or humic soil .... Leskeaceae

4. Plants larger; leaves dull, pinnate or with branches upright and clustered; not in tangled mats 5

5. Leaves long-toothed, dull, closely appressed .... *Thelia hirtella* in Theliaceae

> This is a tree base moss in New England and north, a tree trunk moss farther south.

5. Leaves entire, dull, divergent when moist, loosely appressed when dry; branches tapering and growing away from the substrate .... *Anomodon attenuatus* in Anomodontaceae

Tree Trunks (generally above 0.5 m or so)

Very few species can stand the desiccation encountered on a tree trunk well above the forest floor, but there are some, and for the most part they're not too difficult to identify, at least to genus.

1. Plants with regularly pinnate branching; capsules long-exserted 2
1. Plants randomly branched, not pinnately branched; capsules immersed to long-exserted 3

2. Leaves shiny, falcate-secund, entire .... *Hypnum* in Hypnaceae
2. Leaves dull, closely appressed, long-toothed .... *Thelia hirtella* in Theliaceae

3. Small to large plants; branches growing outward and upward away from the tree trunk; capsules immersed to long-exserted; leaf cells lacking papillae, so leaves are more reflective and shiny 4

3. Small to medium-sized plants; stems and branches forming tangled mats; capsules long-exserted; leaves of most species opaque and dull looking from papillae on cell surfaces 6

4. Stems and branches very flattened; leaves wavy; capsules immersed to barely above leaves...*Neckera pennata* in Neckeraceae

4. Stems and branches round, not flattened; capsules immersed to long-exserted...5

5. Large plants with stems to 4 cm long; leaves to 2 mm long, sometimes with microphyllous branches in leaf axils; capsules immersed to slightly exserted .... Leucodontaceae

5. Small plants with stems 2 cm long or much less; leaves ≈1 mm long; capsules long exserted................................*Platygyrium* and *Pylaisia* in Hypnaceae

6. Small wiry mosses with tiny, often broken leaves and a short, thin costa that is difficult to see ....*Haplohymenium triste* in Anomodontaceae

6. Clearly costate mosses with leaves unbroken, often in floodplains ....
.... Leskeaceae

Waste Places

This category includes road and field edges, gravel parking areas, campsites, gardens, and generally any disturbed, usually sunny site. Most of these species are very good capsule producers, presumably because the ephemeral nature of their substrate forces them to find new locations frequently. This is acrocarp territory, but if the waste place becomes damp and shaded, forest floor species may be encountered, particularly the Hypnaceae and Brachytheciaceae families.

Rock

Most of our rock mosses are dark green to blackish-green acrocarps with immersed capsules, or capsules slightly exserted on a short seta. See Acrocarpous Mosses.

1. Growing on rock; leaves falcate-secund and ecostate, shiny; capsules long-exserted.. *Hypnum* in Hypnaceae

1. Growing on rock with leaves not falcate-secund, costate or not, shiny or not; capsules immersed or long-exserted..2

2. Leaves dull, nonreflective (papillose leaf cells)..3
2. Leaves shiny, reflective (smooth leaf cells)...4

3. Leaf tips whitish, lacking chlorophyll; on acidic rock; capsules immersed at branch tips ....*Hedwigia ciliata* in Hedwigiaceae
  Easily confused with the acrocarpous mosses, this is a common rock moss, and an
            important one to learn early—see family description.

3. Leaf tips green; on calcareous rock; capsules long-exserted; often with tightly packed upright branches mimicking an acrocarpous look ....
.... *Anomodon* in Anomodontaceae

4. Leaf arrangement complanate...5
4. Leaf arrangement not complanate; dendroid plants; on wet, shaded acidic or calcareous rock or soil over rock..
....*Thamnobryum alleghaniense* in Neckeraceae

5. Leaves with a thin costa extending little more than half the leaf length; on calcareous rock or soil over rock....... *Homalia trichomanoides,* in Neckeraceae
5. Leaves ecostate; on damp acidic or calcareous rock or soil over rock ....
.... *Pseudotaxiphyllum elegans* and *Taxiphyllum deplanatum,*
both in Hypnaceae

Dry Open Soil

Generally not pleurocarpous moss territory. See Waste Places in the Acrocarpous section.

FAMILY DESCRIPTIONS

Almost all the pleurocarpous families in our area are in the order Hypnales, with the exception of the taxonomically uncertain Hedwigiaceae, currently residing in its own order, Hedwigiales. Hypnales capsules have 16 outer teeth, with another 16 teeth in an inner ring occasionally present. The similarity in capsules throughout the pleurocarps makes capsule structure less helpful in separating these families than in the acrocarps, with their much greater capsule (and taxonomic) diversity.

Amblystegiaceae

Genera Included *Calliergon, Campyliadelphus, Campylium, Drepanocladus Hygroamblystegium, Hygrohypnum, Leptodictyum, Sanionia, Straminergon, Warnstorfia*

These are generally large well-branched plants of very wet places. Most species are semiaquatic to aquatic at least seasonally. *Campylium* spp. are somewhat drier

An Amblystegiaceae leaf sampler; left to right, *Calliergon cordifolium, Campylium stellatum, Campyliadelphus chrysophyllus, Hygrohypnum ochraceum, Leptodictyum riparium, Sanionia uncinata* (Bry. Eur.)

sited and are small to medium-sized. Some species favor calcareous habitats.

Capsules are cylindrical, curved, constricted below the mouth when dry, with 16 outer teeth that are often light-colored, contrasting with the darker capsules. There are also 16 inner teeth that may appear as a second, inner ring. Setae are long, red to red-yellow.

Leaves are the same on stem and branch, usually lanceolate to acuminate, sometimes ovate-lanceolate, with a long single costa (exceptions are *Campylium hispidulum* and *C. stellatum,* which are ecostate, and *Hygrohypnum ochraceum,* which has a sometimes visible double costa).

Hypnaceae species have similar capsules, but they are drier sited and ecostate. Brachytheciaceae species are costate, as are the Amblystegiaceae, but they have shorter, stocky capsules, and, like the Hypnaceae, are generally drier sited.

This family has ≈300 species worldwide in 23 genera, with 13 genera found in eastern United States.

Here, 12 species in 9 genera are included.

See a more complete discussion of *Sanionia* and *Warnstorfia* on page 229.

Anomodontaceae

Genera Included *Anomodon, Haplohymenium*

Small to large plants with horizontal primary stems with red-brown rhizoids and upright secondary stems lacking paraphyllia or pseudoparaphyllia. Leaves are costate and dull and nonreflective from papillose cells. Capsules are produced infrequently in some species, not at all in others. In some *Anomodon* spp., the densely packed upright secondary stems can mimic the tufted look of acrocarpous species.

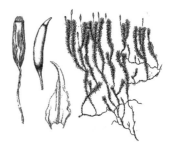

Anomodon viticulosus

This family comprises 4 genera worldwide, with only 2 in our coverage area. In some references these genera are included in Thuidaceae, another family of pleurocarpous mosses with papillose leaf cells. On tree bases or rock.

Brachytheciaceae

Genera Included *Brachythecium, Bryhnia, Bryoandersonia, Pseudoscleropodium, Tomentypnum*

Large, generally well-branched plants with creeping, rarely upright stems, in wet to dry sites. *Brachythecium* spp. are generally the driest sited of the group, followed by the damp soil and rotting wood species *Bryhnia* and *Bryoandersonia* and a wet fen species, *Tomentypnum nitens*.

Leaves are elongate, pointed, costate, often acuminate, spreading to erect, usually not falcate-secund. Many *Brachythecium* spp. and *Tomentypnum nitens* leaves are plicate and erect to

Brachythecium sp. (Bry. Eur.)

spreading. *Bryhnia, Bryoandersonia,* and *Pseudoscleropodium* leaves are concave and closely appressed, at least when dry.

Capsules have 16 teeth, are short and stocky and often dark reddish brown, on a long seta holding the capsules well above the leaves.

Hypnaceae species lack the costa, and have leaves often falcate-secund.

This is a large family with >50 genera and 550 species worldwide. In eastern United States, a dozen or so genera can be found, depending on who is counting, with the *Brachythecium* genus accounting for most of the species.

See a more complete discussion on page 237.

Climaciaceae

Genus Included *Climacium*

Large treelike mosses growing from subterranean stems in swampy places. Stems and branches have a tomentose look from dense paraphyllia.

Capsules erect on a long seta.

Leaves are costate, variably plicate, and shiny. Stem leaves are larger and with less chlorophyll (less green) than the branch leaves.

This family comprises only one genus, and two common species. One is illustrated, both are discussed.

Climacium americanum

Entodontaceae

Genus Included *Entodon*

Medium-sized glossy plants with ecostate, concave leaves tightly appressed to stems (julaceous); the stems sometimes with a partially flattened aspect.

A. J. Grout (1903) said of this genus, "Their brilliant color and flattened habit render this genus easy of recognition. The leaves are so much nearer together than in

Entodon cladorrhizans

most other flattened forms, that one is not likely to put them with forms like plagiothecium." (I should add that *E. seductrix* is not flattened.)

Capsules red-brown, elongate, upright, with 16 teeth.

The family comprises 3 genera and ≈200 species worldwide. In eastern North America, 1 genus and 7 possible species are found, at least 4 of which are rare or uncommon. One species is illustrated here, one more is discussed.

Fontinalaceae

Genera Included *Dichelyma, Fontinalis*

Long (some species measure in feet!), stringy, very dark green aquatic plants anchored to substrate at the base only.

Capsules are on a very short seta, immersed in leaves.

The leaves of *Fontinalis* spp. are ecostate. *Dichelyma*, a

Fontinalis antipyretica (M.V.T.)

costate genus, is likely to be inundated only at high water. *Brachylema*, another costate genus, has a single species in the southeastern U.S. coastal plain.

Four genera occur worldwide, three in northeastern North America.

Hedwigiaceae

Genus Included *Hedwigia*

This moss has some pleurocarpous and some acrocarpous characteristics, so it's a good one to learn early in the game. Growing on rock, it is much branched, with a sprawling, decidedly pleurocarpous look; however, its capsules are formed at the ends of branches, an acrocarpous characteristic. Fortunately, it is an easy moss to learn. It lacks chlorophyll at the ends of leaves, and its cells have bumps (papillae), so they have a glaucous and dull, unreflective look, particularly when dry. Also, its look changes dramatically from wet to dry (see drawings at right and photos on the species page).

The family, in order Hedwigiales, is the only moss keyed here as a pleurocarp that is not in the order Hypnales. Its capsules, immersed in leaves, are kettle-shaped, lack teeth, and are very different from the generally more cylindrical, 16-toothed capsules in order Hypnales.

Hedwigia ciliata
(Bry. Eur.)

H. ciliata is most likely to be confused with the acrocarpous Grimmiaceae family, though *H. ciliata* leaves are ecostate, and Grimmiaceae species all have costate leaves.

H. ciliata is the only representative of this family in eastern North America.

Hylocomiaceae

Genera Included *Hylocomiastrum, Hylocomium, Loeskeobryum, Pleurozium, Rhytidiadelphus, Rhytidium*

This is a family of large (several species >5 cm long), conspicuous, much branched, often pinnate to bi- or even tripinnate, forest floor plants. Leaves are large (>2 mm), shiny, frequently plicate, and lacking a prominent single costa (*Rhytidium rugosum* is the costate exception).

Hypnaceae species may also have large, much branched, variously pinnate, and shiny ecostate leaves. Leaves in Hypnaceae are often falcate-secund, while Hylocomiaceae species have leaves

Hylocomium splendens

that are straight, wrinkled, plicate, or squarrose, but not falcate-secund. Brachytheciaceae species are costate.

This family has 9 possible species in eastern North America; 8 are discussed here, skipping only 1 uncommon calcareous species.

Hypnaceae

Genera Included *Callicladium, Calliergonella, Herzogiella, Hypnum, Platygyrium, Pseudotaxiphyllum, Ptilium, Pylaisia, Taxiphyllum*

Howard Crum (2004) described this as a "family of pleurocarp leftovers," and it is indeed a diverse group, difficult to characterize. Most are large soil and forest floor species, though rock and tree species

Left to right: *Ptilium crista-castrensis, Hypnum imponens, Callicladium haldanianum*

are included. The family members keyed or illustrated in this book have smooth, long-pointed, acuminate leaves without a costa, often shiny. *Hypnum* spp., often the most commonly encountered, have leaves that are long-pointed and falcate-secund.

As with most of the pleurocarpous families, capsules are cylindrical, inclined to horizontal, curved or straight, with 16 teeth.

This huge family has 73 genera and >2500 species worldwide. Of the 15 possible genera in eastern North America, 9 are keyed here.

See a more complete discussion of the *Hypnum* genus on page 258.

Leskeaceae

Genera Included *Leskea, Leskeella*

Small plants (stems <1 mm diameter including leaves) of tree trunks, tree bases, and humic soil, often in rich floodplains. Leaves are costate, small (<1 mm), long-pointed, and entire. *Leskea* leaf cells are papillose and therefore may be dull looking while *Leskeella* leaf cells are smooth, and more likely to look shiny.

Leskea polycarpa (top right) is a tree trunk species, whereas *Leskeella nervosa* (bottom right) can be found on tree trunks, tree bases, and humic soil. *Leskea polycarpa*, true to its species name, commonly produces capsules.

Upper, Bry. Eur.; lower, Bulletin of the Torrey Botanical Club

This small family has just these 2 genera in eastern North America, with a total of 5 possible species; 1 species from each genus is included here.

Leucodontaceae

Genera Included *Forsstroemia, Leucodon*

Large plants with stems ≤4 cm long, growing outward and upward on trunks of trees; branches often julaceous; leaves large, ≤2 mm long, often shiny (not papillose), and generally ecostate, though *Forrstroemia trichomitra* is sometimes costate with a single costa, and sometimes not!

Capsules, on a very short setae and immersed in leaves, are oval to cylindrical with 16 teeth.

Left, *Leucodon brachypus*; right, *Forrstroemia trichomitria*

See Tree Trunk category in the quick look by habitat for separation from other tree trunk pleurocarp families.

Three *Leucodon* spp. occur in our area, and the most common two are discussed. *Forrstroemia* has 2 species in the Northeast; 1 is included here, and the other two are unlikely to be encountered.

Neckeraceae

Genera Included *Homalia, Neckera, Thamnobryum*

All species are large, with stems ≤5 cm long and leaves 2–3 mm long. Leaves are variable. *Thamnobryum* has a dendroid look and leaves with a strong single costa;

Left to right: *Thamnobryum alleghaniense, Neckera pennata, Homalia trichomanoides*

Homalia is complanate, with leaves having a thin indistinct costa extending little more than half the leaf length; *Neckera* is also complanate with no costa and very wavy leaves.

Habitat is as diverse as the leaf structure and arrangement. *Homalia trichomanoides* is found on wet calcareous rock or soil over rock, *Neckera pennata* is a tree trunk species, and *Thamnobryum alleghaniense* is found on shaded wet rocks (calcareous or acidic) along water.

Capsules, all with 16 outer teeth, are on a long seta and held well above the leaves in *Homalia*, on a short seta partially immersed in leaves in *Neckera*, and on a long seta in *Thamnobryum*, though capsules are rarely produced in that species.

The strongly costate, not complanate *Thamnobryum* genus seems out of place here. More than 100 years ago A. J. Grout (1903) wrote that including

Thamnobryum in this family "is an arrangement with which I am compelled to thoroughly disagree."

One species per genus is included here. These are the only species in this family likely to be encountered in northeastern North America.

Plagiotheciaceae

Genus Included *Plagiothecium*

Glossy, usually complanate plants with ecostate leaves occasionally somewhat asymmetrical as a result of being twisted around the stem to a flattened (complanate) position. Species in this family favor wet mineral soil or seepy ledges.

Capsules are cylindrical, erect to inclined, with 16 teeth.

The only complanate, ecostate Neckeraceae species is *Neckera pennata*, a tree trunk species with distinctive, wavy leaves.

The family has only 1 genus and 5 species in northeastern North America, and the 3 most common and conspicuous are included here.

Plagiothecium denticulatum
(Bry. Eur.)

Sematophyllaceae

Genus Included *Pylaisiadelpha*

Small, glossy, yellow-green plants with stems 2–5 cm long, and complanate, ecostate, falcate-secund leaves. Habitat is conifer forest floor, often at high elevations.

Capsules are cylindrical and with 16 teeth.

Hypnum species in Hypnaceae are generally larger, but very similar.

Two genera occur in eastern North America. Two are generally southeastern in distribution. *P. recurvans*, the species keyed and illustrated here, is the species most likely to be encountered in our range.

P. recurvans

Theliaceae

Genera Included *Myurella*, *Thelia*

Plants are medium-sized with stems 3–5 cm long, glaucous to yellow-green; leaves small (<1 mm long) and tightly packed (julaceous). *Thelia hirtellla* (the species included here), a tree base and trunk species, is regularly pinnate with long-toothed, costate leaves and cells with papillae, therefore looking dull, or at least not

Left, *Myurella sibirica*; right, *Thelia hirtella* (Ic.)

shiny. *Myurella* species are more randomly branched, are found on soil over calcareous rock, are ecostate, and have leaf cells with papillae or not.

Capsules are ovoid-cylindrical, have 16 teeth, and are held well above the branches. There are 5 species in the family in eastern North America.

Thuidiaceae

Genera Included *Abietinella, Helodium, Thuidium*

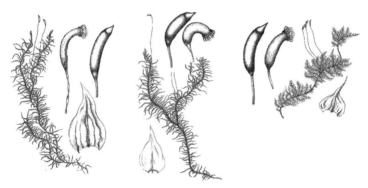

Left to right: *Abietinella abietina, Helodium blandowii, Thuidium delicatulum*

Large plants (some species ≥10 cm long), randomly to singly or bipinnately branched, often with upright growing stems with paraphyllia; leaf cells papillose and therefore dull and nonreflective. Habitat can be damp shaded soil, calcareous rock, tree trunk, or tree base.

Capsules, with 16 teeth, are cylindrical, often inclined, and on long setae.

This family (and many others) is not particularly stable, but as of this writing, 33 genera and >500 species are listed worldwide.

Pleurocarpous Species Accounts
Presented alphabetically by family

Above and right, *C. cordifolium* from a muddy pond margin at West Branch Pond Camps, Shawtown, ME; middle right, capsules of *C. cordifolium* from an herbarium collection; lower right, *Straminergon stramineum* from Crystal Bog, a rich fen in Sherman, ME

Description These very similar species are large, yellow-green plants with stems ≤15 cm long, sparse branching, and leaves well spaced and wide spreading, usually more imbricate at the stem tips. The stem is distinctly ridged from leaf decurrencies. Leaves are 2–3 mm long, costate (hard to see because of striations), concave, mildly striate, and sometimes with cucullate tips. When dry and twisted, the blunt tips appear pointed. Capsules are frequently present.

Similar Species *S. stramineum* (lower right) is very similar but has diagnostic tomentum at the tips of at least some usually older leaves lower on the stem. Its distribution is much more northern, reaching south only into northern New England and northern NY, and it is found in rich fens. See also *C. giganteum* and *Calliergonella* (Hypnaceae) Similar Species.

Range and Habitat *C. cordifolium* is found throughout our range, and south as far as North Carolina in higher terrain. Semiaquatic, it's found in swamps, cattail marshes, and lake margins.

Name *calli* (G) = beauty, plus *ergo* (G) = work, for a beautifully made moss; *cord* (L) = heart, plus *folium* (L) = leaf, for the heart-shaped leaf base, and *stramen* (L) = straw, for the color.

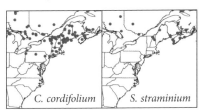

C. cordifolium	*S. straminium*

Amblystegiaceae　　　　　　　　　　　　　　　　Ca for *S. stramineum*

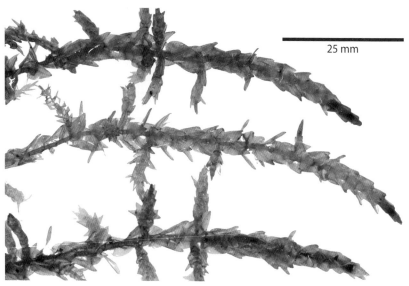

At edge of water on limestone, mixed with *Calliergonella cuspidata*, Symmington Quarry, Rockland, ME

Description A large yellow-green to green plant with stems ≤20 cm long, ascending, with pinnate branching; leaves 2–3 mm long, costate, concave, very striate, cucullate, well spaced, and wide spreading, except at the stem tips, where they are more imbricate. Capsules uncommon.

Similar Species *C. cordifolium* is only slightly smaller, is much less branched, and its leaves are less striate and less wide than long, compared with *C. giganteum*, which has leaves very striate and almost as wide as long. Also, *C. cordifolium* is a good capsule producer, while *C. giganteum* rarely produces them. See also *C. cordifolium* and *Calliergonella* (Hypnaceae) Similar Species.

Range and Habitat Primarily in the northern part of our range, south to Pennsylvania. Semiaquatic in fens, shallow vernal pools, cattail marshes, and lake margins, probably preferring calcareous habitats.

Name *Giganteum* rather obviously refers to the size of this large moss, though it's not much larger than *C. cordifolium*.

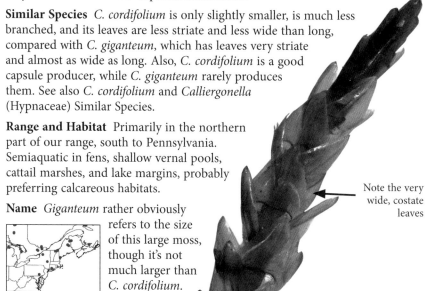

Note the very wide, costate leaves

C. giganteum

Above and right, on a Native
American shell midden,
Damariscotta, ME

Description In small, light to dark
green tufts, this is the smallest of the
Campylium/ Campyliadelphus group,
with leaves <1 mm long, wide
spreading to squarrose, ecostate
(actually double and microscopic costa),
smooth, and not much contorted when
dry.

Similar Species See the other
Campylium and *Campyliadelphus* species.
Substrate preference and size is probably
the best way to differentiate the species.

Range and Habitat Throughout our
range on roots, decorticate logs, and
decaying wood in moist and shady
places, with some preference for enriched

C. hispidulum

sites.

Name *campyl*
(G) = curved, for the
capsules; *hispid*
(L) = bristly, a pretty good description for all these
squarrose *Campylium* species.

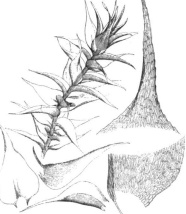

Campylium hispidulum (From Sulliv. "Icones")

Top and above right, from a gravel pit in Augusta, ME; above left, capsules from an herbarium sample collected at Crystal Bog, Aroostook County, ME

Description Medium-sized with dense, upright, golden-green stems ≈4–5 cm long and little branched; leaves wide spreading to squarrose, somewhat distant, >2 mm long, with long-acuminate, channeled (margins rolled inward) tips, ecostate, plicate, and twisted when dry. Capsules produced occasionally.

Similar Species All the *Campylium* spp. have this squarrose look, with twisted acuminate leaf tips, but this is the largest and the only one likely to be found in wet calcareous habitats.

C. stellatum

Campylium stellatum (From Bry. Eur.)

Range and Habitat Throughout our range, a dedicated calciphile of rich fens, meadows, and bogs.

Name *stella* (L) = star, for the squarrose leaves.

Top and lower left, Eagle Hill, Steuben, ME; lower right, from a shell midden, beside a stem of the more julaceous *Anomodon rostratus*, Damariscotta, ME

Description Medium-sized irregularly branched plants yellow-green to brownish green, with stems ≤5 cm long; leaves ≈1.5 mm long, costate, wide spreading to squarrose, with acuminate tips; capsules smooth, not ridged.

Similar Species See *Campylium* spp. This is the only costate member of the Campylium/Campyliadelphus group, but that character can be tricky, given the small size of the leaves, their plicate nature when dry, and variability in costa length. Both the other common *Campylium* spp. show stronger preference for calcareous substrate, while this species is more of a generalist. See also *Herzogiella striatella*, a very similar, but ecostate species.

Range and Habitat Throughout our range on rock or gravelly soil in wet places. Occasionally on rotting wood. This species may be more of a substrate generalist than the other two *Campylium* spp., which seem to require some enrichment.

Name *Campyliadelphus* is for a *Campylium* sibling; *chrys* (G) = gold, and *phyllum* (G) = leaf, for the yellow-green color of this species.

C. chrysophyllus

Amblystegiaceae

Winslow Ledges, a calcareous ledge along the Kennebec River, Winslow, ME

Description A small, wiry, yellow-green moss, with small
(≈1.5 mm), strongly costate (dark costa), acuminate
leaves spaced along the stem so that the red-brown
stem is occasionally visible. The leaves typically spread
outward at the base and curve inward to parallel with
the stem. In very old parts of the stem, the leaves
erode away and the costae remain (below right).
According to Janice Glime (1993), this is typical for the
genus. Capsules are infrequently produced.

Similar Species [*H. tenax*], a close relative, has narrower,
sharper leaves with a yellow costa. Also, [*H. tenax*]
leaves diverge from the stem at the leaf tips rather than
curving inward as with this species. *H. fluviatile* is a
calciphile whereas [*H. tenax*] may be found in both
calcareous or noncalcareous situations.

Range and Habitat From Maine and New Hampshire
west to Minnesota and south to North Carolina,

H. fluviatile

occasionally to frequently on submerged
calcareous rock.

Name *hygro* (G) = moist, and *Amblystegium* is a moss
genus name from *ambly* (G) = blunt and *stego* (G) = a roof
or cover, referring to the blunt operculum on the
infrequently produced capsules; *fluviatil* (L) = river. So we
have a river moss with blunt capsules.

Amblystegiaceae   Ca

Above, collected by Cloe Chunn in rich woods, Northport, ME; right, a collection from Phippsburg, ME

Description In streams in loose, floating, yellow-green to brownish mats of branched, usually parallel stems ≤8 cm long; leaves broad, acuminate, falcate-secund, concave, ≈1.5–2.6 mm long, with a costa varying from short, double, and hard to see, to forked, extending to half the leaf. Leaf tips are usually twisted and leaves are more or less striate when dry. This species, like most aquatics, is extremely variable. It generally looks like a drowned *Hypnum*. Capsules uncommon.

Similar Species *Hypnum lindbergii* is similar and will be found next to brooks but is usually not immersed, and its leaves are shiny and not striate when dry. *Drepanocladus* spp. are more obviously costate. Other species of *Hygrohypnum* might also be found in our coverage area. Two are somewhat common: [*H. luridum*], similar to this but smaller, with leaves <1.5 mm long, and [*H. duriusculum*], with leaves ≈1.5 mm long, divergent, and almost as wide as long.

H. ochraceum

Range and Habitat Newfoundland to Minnesota south to North Carolina, on sand and rocks in small, quick-moving streams, brooks, or waterfalls, possibly favoring high-elevation sites. Not found in calcareous sites.

Name *hygro* (G) = moist, *Hypnum*, for that genus, and *ochraceus* (L) = yellow with a brownish tint, for the color of the plants.

Amblystegiaceae

25 mm

1 mm

Above, from a brook in calcareous woods in Ravena, NY; below right, capsules from an herbarium collection

Description Long, irregularly branched stems ≤10 cm long, forming loose, tangled mats near or in water; leaves somewhat complanate, wide spreading, sparse, 2–4 mm long, costate (to 3/4 of leaf), entire, tapering to a long, slender tip. As with most of our (occasionally) aquatic species, there is much variation, and many varieties and forms have been described. Capsules are occasionally produced.

Similar Species Hook Mosses (p. 229) share the habitat and general form, but they have more curved stem tips and lack the wide-spreading leaves. *Fontinalis* spp. are ecostate, and *Campylium* spp., members of the same family, have a similar look but are not aquatic.

Range and Habitat Throughout North America, forming wet mats in or next to water, particularly in hardwood swamps subject to seasonal flooding. May be found in habitats that are calcareous as well as noncalcareous.

Name *lept* (G) = slender, *dicty* (G) = a net, referring to the pattern of cells, which in this species are slender (for this family); *riparius* (L) = belonging to a stream, for the habitat.

L. riparium

Amblystegiaceae

Key to Hook Mosses (formerly genus *Drepanocladus*)

How we got here: These are pleurocarps, not complanate, irregularly branched, with long-acuminate, costate leaves (much) more than 1 mm long, not squarrose, strongly falcate-secund, especially at stem and branch tips.

The Hook Moss species resemble a large, exaggerated *Hypnum* sp., but with a pronounced hook at the end of the stem, and without well-organized pinnate branching. Their size and the presence of a costa will separate them from the *Hypnum* spp. All but *Sanionia uncinata* are bog or fen species.

The hook moss species discussed here, all formerly in the genus *Drepanocladus*, are mostly now in other genera, but because of their morphological affinities, I key them together. Aquatic species are often highly variable, and this group is no exception. Hook mosses can be very difficult to key to species with or without a compound microscope, so you should be pleased to have made it this far with a hand lens. The following information is offered to help you get to species, though the level of confidence might be low.

The following are all serrulate at least at the leaf tip, though the serrulations typically comprise only one cell and can be very hard to see with a hand lens:

A. Stem and usually branch leaves plicate; plants green or yellowish green; in shaded coniferous woods on bases of trees, humus, rotten logs, stumps, and rocks, or occasionally in wet meadows and drainage ditches. Never submerged and never red to purple.........................*Sanionia uncinata*

A. Stem and branch leaves smooth; plants often tinged with red or purple, seasonally submerged...B

B. Costae extending 3/4 or more the length of the leaves; often submerged in bogs, swamps, meadows, wet depressions, and drainage ditches..............
.... *Warnstorfia exannulata*

> See also *Dichelyma capillaceum* in Fontinalaceae, which may key here.

B. Costae extending to middle of leaves, rarely beyond; in bogs, swamps, wet depressions, and at margins of streams and lakes. Very similar to *W. exannulata*...[*Warnstorfia fluitans*]

Three Entire Species An entire species looking like, and found in habitats described for, *W. exannulata* and [*W. fluitans*] would probably be [*Drepanocladus aduncus*], our sole remaining member of the original genus. Two formerly genus *Drepanocladus* Hook Mosses not illustrated here may be expected in calcareous fens. [*Scorpidium revolvens*] and [*Hamatocaulis vernicosus*] are both entire. [*S. revolvens*] is red to black, with a central strand in the stem, whereas [*H. vernicosus*] is green and without the central strand. Both are considered rare, probably because their habitat is so uncommon.

Amblystegiaceae   Continued on next page

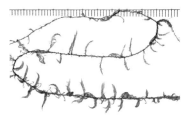

W. exannulata, showing how stringy aquatics can be. Hash marks at intervals of 1 mm

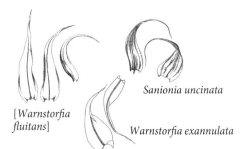

[*Warnstorfia fluitans*]

Sanionia uncinata

Warnstorfia exannulata

Photos by Hermann Schachner,
courtesy of Wikimedia Commons

Description Green to yellowish-brown plants with prostrate or upright, irregularly to subpinnately branched stems. The costate leaves, 2.5–4 mm long, are noticeably plicate and long-acuminate with long, curved tips woven around the stem. Monoicous and usually a good capsule producer. See also drawing of leaf on previous page.

Similar Species This is the only hook moss that lives in anything other than a very wet, probably seasonally inundated, area. It has plicate leaves versus smooth for *W. exannulata* and [*W. fluitans*]. *Hypnum* species are smaller, usually more regularly branched, with leaves ecostate and not plicate. *Brachythecium* species lack the exaggerated falcate-secund leaves. See "Key to Hook Mosses" (p. 229).

S. uncinata

Range and Habitat Throughout the Northeast, south to Pennsylvania in moist forests on decaying wood, humus, and tree bases, and among grasses in wet meadows.

Name The genus is named after the German botanist C. G. Sanio (1832–1891); *uncinatus* (L) = barbed, for the hook-shaped branches.

Amblystegiaceae

Top, photo by Hermann Schachner, courtesy of Wikimedia Commons; middle and bottom right, photos by the author, Aurora, ME

Description Forms dense colonies of yellow-green to dark green, often brown to purple below, infrequently branched to subpinnately branched stems to several cm long. Leaves are shiny, not plicate when wet, mildly plicate to not when dry, long-acuminate, concave, strongly falcate-secund, with a costa extending to 3/4 the length of the leaf. The hooked tip is diagnostic for the group. Crum and Anderson (1981) said, "This species has been divided into a fantastic number of varieties, forms, and subforms," so don't be surprised if you run into some variation from the photos shown here. Note the distinctive capsule.

1 mm

Similar Species [*W. fluitans*], similar in form and habitat, has a costa extending only half the leaf length, and cellular differences along the base of the leaf. *Hygrohypnum ochraceum* usually has no visible costa. *Sanionia uncinata* has plicate leaves. *Dichelyma capillaceum* (Fontinalaceae) has the "hook" look with smooth leaves but is more likely to be in a seasonally inundated or stranded situation, and it has short setae. See "Key to Hook Mosses" (p. 229).

Range and Habitat A northern species, found from Greenland to Alaska, and

W. exannulata   *W. fluitans*

in our range, south to New England. In streams, bogs, swamps, and ponds.

Name *Warnstorfia* is for Carl Friedrich Warnstorf (1837–1921), German *Sphagnum* specialist; *ex* (L) = comparable to, and *annulus* (L) = a ring, for the very curved leaves.

Amblystegiaceae

On hardwood tree base, rich woods, Hancock County, ME

Description A very distinctive tree base moss that grows out away from the tree trunk and then droops with tapering branches. The leaves are blunt-ended, apiculate, costate, and covered with papillae (as are all species in this genus), giving this species the dull, nonreflective look shown clearly in the "dry" photo (right). When moist, the branches have a complanate look, as shown lower right. Capsules are rare.

Similar Species [*A. rugelii*] is similar, sharing the tree base habit, but its leaves are blunt, not apiculate, its branches are not attenuate, and it's uncommon south of northern New England and southern Ontario. *Leucodon* spp. (p. 277) also grow outward from bark in similar habitats, but they grow out and upward and have pointed leaves. *Neckera pennata* (p. 280) is more complanate, grows down, then outward, and has larger, shiny, undulate leaves. *Porella* liverworts are often found in the same habitats. They grow away from the tree trunk in flat sprays with rounded, tightly packed, two-ranked leaves. All these similar species other than [*A. rugelii*] grow higher on the trunk than *A. attenuatus*.

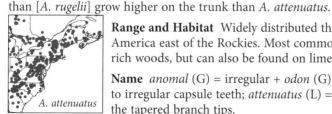

Range and Habitat Widely distributed throughout North America east of the Rockies. Most common on tree bases in rich woods, but can also be found on limestone rock.

Name *anomal* (G) = irregular + *odon* (G) = teeth, referring to irregular capsule teeth; *attenuatus* (L) = diminished, for the tapered branch tips.

Anomodontaceae

A. *rostratus* on limestone, Ravena, NY

Description Forms dense yellow-green mats with crowded, short, erect, slender, terete branches rising straight up from a creeping main stem. Branches appear randomly arranged along the stem, short, and tiny, ≈0.25 mm across, including the tightly appressed leaves. The lower parts at the base of the carpet are yellow to rust colored. Densely packed leaves are dull because of multiple papillae per cell (true of all species in this genus). Leaves have a prominent bulging costa (yellow?), long, smooth, hyaline hair points, and revolute margins, and they are julaceous when dry and erect-spreading when wet. Dioicous and an infrequent producer of capsules measuring 1–1.5 mm, oval, smooth, turning brown when mature.

full size

Anomodon rostratus (From Bry. Eur.)

Similar Species The hyaline hair points separate this from the other *Anomodon* spp. Also, this is the smallest *Anomodon* species, easily separated by size and color from *A. viticulosus*, another rock *Anomodon* species, and the largest.

A. *rostratus*

Range and Habitat Common throughout eastern North America, usually on shaded calcareous rock and cliff crevices, occasionally on bases of trees, and on humic soil.

Name *rostratum* (L) = beaked, for the operculum shape.

A. viticulosis on limestone, Ravena, NY. Note the smaller and more yellow *A. rostratus* in the lower part of the photo.

Description The largest and most robust of the *Anomodon* genus, with branches ≤10 cm long forming deep mats. Green to yellowish with the underparts of the mat more yellowish. Leaves >2 mm, with strong (bulging? yellow?) costa and blunt apiculate tips, are wide spreading when wet (right), to somewhat contorted and incurved when dry (above). The pleuripapillose leaf cells cause a dull appearance (as with all species in this genus). Capsules are not found in North America.

Similar Species *A. attenuatus* has similar but smaller leaves and sparser branching, grows out and down versus out and up for this species, and is common on tree bases as well as on rock. *A. rostratus* is the smallest of the threesome with straight, upright branches in tight cushions.

A. viticulosis

Range and Habitat A calciphile, found primarily on limestone. Reported from the Canadian Maritimes south to Tennessee.

Name *vitis* (L) = vine and *ulus* (L) = little, for the miniature vine it resembles.

Anomodontaceae   Ca

Above, dry and stringy; below, hydrated, Ricketts Glen State Park, Benton, PA

Description Small, green to yellow, dull, stringy, in thin, tangled mats; irregular branching; leaves appressed when dry (above), wide-spreading when wet (right), <1 mm long, usually broken except near branch tips; leaf cells with papillae, leading to the dull appearance, costate, but costa is thin, short, extending approximately to midleaf, and usually not visible with a hand lens. Dioicous and not known to produce sporophytes in North America.

Note the scale; the leaves are very small, and it is hard to see that they are broken unless the sample is wet and leaves diverge from the stem.

Similar Species *H. triste* resembles the *Anomodon* spp., which are also dull from papillae and frequently found on tree trunks; however, *Anomodon* spp. are larger and don't have the distinctive broken leaf tips.

Range and Habitat Mostly in cool, mature, hardwood forests, possible throughout our range, but uncommon in northern New England and eastern Canada, more likely from upstate New York and south down the spine of the Appalachian Mountains.

Compound microscope ≈50×

H. triste

Name *haplos* (G) = single (or simple), and *hymen* (G) = a membrane, referring to a simple inner peristome (a microscopic capsule characteristic); *trist* (L) = sad, but is used here in reference to the dull appearance.

Anomodontaceae

Brachythecium

How we got here: The genus, with 15 species within our range, is characterized by its pleurocarpous form; soil or tree base habitat; irregularly pinnate branching; costate leaves with costa frequently stopping well short of the leaf tip, leaves acute to acuminate, <7× as long as broad, frequently plicate and with twisted tips, appressed to upright to spreading, but not squarrose; stem leaves frequently larger and more plicate than branch leaves; few or no rhizoids on the stem; capsules are dark and stocky.

The *Brachythecium* genus is one of the more difficult to identify to species, relying almost exclusively on microscopic characteristics that are frequently difficult for the seasoned professional to assess. In *Maine Mosses*, volume 2 (2014) Bruce Allen says, "The genus has some good taxonomic characters, but there is a predominance of character states that are neither abruptly distinct nor intuitively obvious."

B. velutinum, with the somewhat falcate-secund leaves that are unusual for the genus

It is a good day if a beginning bryologist is able to identify these plants to genus. Descriptions of three of the larger and more common species are presented here to help give you the "look and feel" of the group. The key below can help you move farther into the group, but identification to species without use of a technical manual should be considered highly tentative.

Brachythecium Key

A. Plants small, leaves seldom reaching 1.5 mm long; setae rough .... B
A. Plants large, leaves usually >1.5 mm long and reaching 3 mm; setae rough or smooth.. C

B. Leaves falcate-secund; stem leaves short-decurrent, margins plane or recurved at base; in rather dry habitats on soil, often over rock, and on bases of trees..[*B. velutinum*]
B. Leaves straight or nearly so; stem leaves long-decurrent, margins recurved to leaf middle or above; in dry woods on bases of trees, rotten logs and stumps, and humus over rocks...................................[*B. reflexum*]

C. Leaves smooth or plicate, stem leaves long-decurrent; setae rough; occurring in wet habitats on soil, rocks, and logs in and beside creeks, rivers, springs, and seepy places...................................... *B. rivulare* (p. 239)
C. Leaves plicate, stem leaves short-decurrent; setae smooth or rough; in dry to mesic disturbed habitats on rock, soil, humus, rotten stumps and logs, and bases of trees.. D

D. Leaves strongly plicate; setae smooth to roughened; branches not complanate..*B. campestre* (p. 238)
D. Leaves mildly plicate; setae rough; branches often somewhat complanate ....*B. rutabulum* (p. 240)

Along the Tarr's Mountain Trail, Arrowsic, ME

Description A medium to large moss forming extensive bright light green to yellowish-green, loose mats with often upright stems ≤5 cm long. Stem and branch leaves are 2–3 mm long, wide spreading, acuminate, long-pointed, costate, plicate, and with very small (microscopic?) teeth in the upper half. Branching is irregular and the plants have an unkempt look. The yellow to reddish seta is smooth to roughened, 1–2 cm long. Autoicous (monoicous), and a frequent producer of capsules that are dark, short, stocky, 2–3 mm long.

Similar Species The *Hypnum* species all have falcate-secund leaf tips and their leaves are neither plicate nor costate. See *B. rivulare* for differences from that species. *B. rutabulum* shares similar habitat, is frequently in fruit (monoicous), but its leaves are only slightly plicate, have twisted tips, are finely toothed all around versus not toothed, or toothed only in the upper half for this species, and *B. rutabulum* setae are rough. Most earlier references will show what we now call *B. campestre* as *B. salebrosum*.

Range and Habitat Throughout our coverage area, south to North Carolina in tree crotches, on tree bases, logs, humus, and lawns in shaded environments. Also occasionally on rock. Common in shaded, disturbed places that are damp, but much drier than is common for *B. rivulare*.

B. campestre

Name *brachy* (G) = short, and *thec* (G) = a case, for the short, stout spore capsule; *campestris* (L) = growing in a field, possibly for the relatively dry habitat.

Brachytheciaceae

West Branch Pond Camps, Shawtown Township, ME, with frogs and Rolling Rock beer bottle for scale

Description Whitish green to yellowish green, in large soft mats with upright, almost dendroid, irregularly branched secondary stems ≈1.5 mm wide rising from horizontal creeping main stems. Main stem leaves, 2–2.5 mm long, are broad, blunt-pointed (see right), very decurrent, and sparse, while the secondary stem leaves, 1–1.5 mm long, are lanceolate to acuminate, and quite plicate. Leaves are costate with costa extending 2/3–3/4 leaf. Leaves glossy wet or dry (no papillae). Capsules, as shown on right, are 2–3 mm long, with a rough, red-brown seta. Dioicous, and not a good capsule producer.

Similar Species *B. campestre* favors drier habitats, has more long-pointed branch leaves, is much more likely to have capsules, and has a smooth, yellowish to red seta.

Brachythecium rivulare × 1; leaf and capsules × 10.

From Grout 1903

B. rivulare

Range and Habitat Throughout our range, south to North Carolina in wet, seepy places, along stream banks, or in intermittent streambeds on gravel or rocks. Can grow submerged, becoming elongated with few branches and leaves more distant.

Name *rivulus* (L) = stream, for the habitat.

Brachytheciaceae

On a cellar-hole wall, Arrowsic, ME

Description A large moss forming extensive, bright, light green to yellowish-green loose mats with stems ≤5 cm long. Branching is pinnate and ascending. Leaves ≤3 mm long are strongly costate, with the costa extending 3/4 of the leaf, somewhat plicate, and lightly toothed all around. See *B. campestre* for the look of the short, stocky, fresh capsules. The dry capsule shown here is typical for both species, except that the seta for this species is rough (bumpy) as shown.

rough seta

2 mm

Similar Species This and *B. campestre* are very similar, as you might tell from the almost duplicate descriptions. *B. rutabulum* is the larger of the two and has leaves toothed all around, slightly plicate, and with twisted tips, versus toothed in the upper half only, very plicate, and with tips little twisted for *B. campestre*. Seta is smooth to roughened for *B. campestre*, versus papillose for this species.

B. rutabulum

Range and Habitat Throughout our range, south to the Mason-Dixon Line, in shaded sites on soil over rock, tree bases, decaying logs, and lawns. In the northern part of the range, this species seems to favor rich, high pH sites such as the cellar hole where the photo above was taken.

Name *rutabulum* (L) = fire poker or penis, for the shape of the capsule.

Brachytheciaceae

On rotting log in seasonally inundated streambed,
Black Mountain Trail near Red River Camps,
Township T15 R9, ME

Description A bright green, midsized moss
with branches ≤6 cm long (usually less) in
loosely tangled mats with irregular upright
branching; leaves small, 1–1.5 mm long,
almost as wide as long, decurrent, concave,
costate with costa stopping short of the leaf
tip, upright to open when wet (see right), to
slightly more appressed and twisted when dry. Grout (1924) says the leaf tips
are twisted half a turn to the right when dry. Capsules are occasionally
produced, short, 3–3.5 mm long, stocky, *Brachythecium*-like, and with a rough
seta.

Similar Species *Brachythecium rivulare* shares the wet habitat but is larger, with
leaves somewhat plicate and without twisted tips, compared with *Bryhnia
novae-angliae*, with smooth leaves and twisted tips. See also *Hylocomiastrum
umbratum*.

Range and Habitat Throughout northeastern North
America, south to Alabama and Georgia in shady, wet
places on soil and logs, and along slow-
moving streams.

Name *Bryhnia* is for Norwegian
bryologist Nils Bryhn; *novae-angliae* is for
New England, which may indicate its
center of distribution in North America.

B. novae-angliae

2 mm

Brachytheciaceae

All photos, *B. illecebra* on a damp log over a stream, Ravena, NY

Description Tightly packed, yellow-green to brownish, shiny, julaceous, upright, branched stems create a tightly woven carpet effect. Leaves are costate, very concave, imbricate when dry, to moderately spreading when wet, very broad—almost as wide as long, and with an abruptly acuminate tip. Capsules are 2–3 mm long with a tapered neck, on a red to red-orange seta ≤25 mm long.

Similar Species *Entodon seductrix* or [*E. cladorrhizans*] may be similar, but both are ecostate. See also *Pseudoscleropodium purim*.

Range and Habitat From Connecticut and New York south in forests on damp soil or wet logs, probably more common in calcareous sites.

B. illecebra

Name *bryum* (G) = moss, and *andersonia* is for Lewis Anderson, bryologist and coauthor with Howard Crum of *Mosses of Eastern North America*; *illici* (L) = enticing or alluring. The specific name seems to have originated with Johannes Hedwig, who named the species *Hypnum illecebrum* in 1801. He must have found this species quite compelling.

In a shaded, grassy site, Brooklin, ME

Description A large, light to dark green moss with green stems often >10 cm long; leaves concave, overlapping, ≈2 mm long, oval, with twisted, apiculate tips, and costa approximately half the length of the leaf; branching varies from irregularly pinnate to pinnate. Dioicous with capsules unknown in North America.

Similar Species *Pleurozium schreberi*, similar in size and look, lacks twisted leaf tips and has red stems, versus green for *P. purim*. *Calliergonella cuspidata* also has apiculate leaf tips but is ecostate, its branch leaves are much more wide spreading, and its habitat is different. *Bryoandersonia illecebra*, with very similar leaves, is smaller and with less organized branching.

Range and Habitat This is a very common moss in Europe that has become well established, possibly invasive, in the Pacific Northwest and has been found recently at several sites within our range. While it is currently uncommon in the Northeast, it is included here because of its apparent potential for range expansion. Historically used as packing material, this moss has traveled the world.

Name *pseud* (G) = false, and *Scleropodium* is for another moss genus in which it was once included but was later removed; *purus* (L) = clean or pure, probably referring to traditional European use of the species as clean bedding for worms.

Brachytheciaceae

10 mm

5 mm

Top and above, from a Maine herbarium collection; color is unreliable

Description A large (5–10 cm tall), long and feathery, upright, pinnately branched, yellow to yellow-brown plant covered with matted red-brown rhizoids (tomentum), as shown in the lower photo. Leaves are long-lanceolate, plicate, costate (perhaps obscured by plications), 3–4 mm long, with tomentum on the backs of the costae.

Similar Species [*Cratoneuron filicinum*], another large upright (or not) pleurocarpous calciphile, also has dense tomentum on the stems and the back of the costae, but its leaves are much smaller (1–1.8 mm), less lance-shaped, not erect, and not deeply plicate.

Range and Habitat This is primarily a northern tundra species, but it can be found south to Connecticut, Pennsylvania, and Michigan in calcareous fens and *Thuja* swamps.

T. nitens

Name *tomentum* (L) = wool or hair, for a hairy-stemmed *Hypnum* species; *nitens* (L) = shining, for the bright yellow of this species. The Hypnaceae family was once more inclusive than it is now, and this species does not closely resemble species in the modern *Hypnum* genus with their ecostate, falcate-secund, acuminate leaves. Species in the current Brachytheciaceae family were once included in Hypnaceae.

C. americanum, on a shell midden, Damariscotta, ME; inset, capsule ≈5 mm long

Description This is a large, treelike moss ≈5–8 cm tall, shown to the right at slightly larger than life. The apparently bare stalk (stem leaves lack chlorophyll) rising from a horizontal stem, combined with the dendroid look, makes this pair of species an easy field ID. Leaves are 2–3 mm long, shiny, costate, plicate. Capsules, rarely available, are 1.5–3.0 mm long for [*C. dendroides*] and 4–6 mm long for *C. americanum*; however, if lacking capsules, these species are almost impossible to separate without a compound microscope. Upper leaf cells are longer than 5:1 for [*C. dendroides*], and shorter than 5:1 for *C. americanum*.

C. americanum, ≈1.25×

Similar Species *Thamnobryum alleghaniense* has more of a "blown to one side" look, with branches originating in one plane, not all around the stem as with these *Climacium* spp., and it is more inclined, that is, less upright, and has dull, smooth leaves, versus shiny and plicate for these species.

Range and Habitat [*C. dendroides*] is more common in the North, *C. americanum* in the South, with *C. americanum* ranging as far north as midcoast Maine (see maps), and both species are possible throughout most of the range covered by this book. Rich woods.

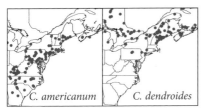

Name *climacium* is from a Greek word for ladder, referring to a microscopic capsule characteristic; *dendroides* is for the treelike form, and *americanum* separates that species from *dendroides*, a species originally described in Europe. Common name: tree moss.

Climaciaceae

On a suburban wall, humic soil over mortar or rock, full sun, Philadelphia, PA

Description Wide, tangled, glossy, yellow to reddish-green mats; stems round and julaceous ≤5 cm long with subpinnate branching; leaves crowded, imbricate, ovate ≈1–2 mm long, and ecostate (actually a microscopic, short, double costa). Monoicous with red-brown erect capsules and setae, capsules 2–3.5 mm long, 16 teeth.

Similar Species [*E. cladorrhizans*] is similar but with flattened branches. *Bryoandersonia illecebra* has a similar look, but its stems are probably shorter, it has costate leaves, and it is found in wetter, often enriched sites.

Range and Habitat More common in the southern part of our range, and south to Florida. On rotten wood, tree bases, and humic soil.

Name *ento* (G) = within, and *odont* (G) = tooth, referring to the peristome teeth hidden below the mouth of the capsule. The specific epithet was originally applied by Johannes Hedwig (1730–1799), author of *Species Muscorum Frondosorum* (1801), the starting point for all moss nomenclature except the Sphagnaceae. With regard to the specific epithet, M. Hedwig clearly spent too many years in the lab.

E. seductrix

Entodontaceae

Seasonally inundated floodplain along the Saco River in Fryeburg, ME

Description In frequently extensive soft mats varying from glossy green to brown, with generally unorganized branching. The leaves, ≈3–7 mm long, are long-acuminate, costate, concave to keeled, not plicate, falcate-secund, and with an excurrent costa and fine tip. Capsules (bottom right) are short-stalked with surrounding leaves typically reaching or covering the capsule base.

Similar Species The hook mosses (*Sanionia uncinata*, *Warnstorfia exannulata*, and [*W. fluitans*] are similar, but all three have exserted capsules, and *S. uncinata* has plicate, not smooth, leaves.

Range and Habitat Throughout the Northeast and south to North Carolina and Tennessee in bogs and swamps, and at lake and stream margins. The sample shown here is high and dry except in spring floods.

D. capillaceum

Name *dicha* (G) = split in two, plus *elemos* (G) = a case or quiver for a split calyptra; *capillus* (L) = hair, plus *aceus* (L) = pertaining to, for the fine, hairlike leaf tip.

Fontinalaceae

25 mm

5 mm

Collected in a small, flowing stream in Putney, VT

Description This is a large aquatic moss, often more than a meter long, triangular in cross section, and with softly folded, ecostate leaves. Green to yellow-brown; with stems red, largely unbranched, though sometimes with a few short branches, anchored at one end; the plants flowing in large, frequently tangled masses in a stream. Leaves are 4–7.5 mm long in three ranks. As they age, samples take on the smell of rotten fish from decaying invertebrates.

Similar Species *F. dalecarlica* and *F. novae-angliae* are smaller, with smaller leaves, and without the folded leaves and the triangular cross section. *F. novae-angliae* is more likely to be in a tangle in still water, the other two more likely found in moving water.

Range and Habitat Reported throughout northern North America; in the Northeast it is found south to Michigan and Pennsylvania in swift-moving streams, sometimes in stream pools.

F. antipyretica

Name *fontinalis* (L) = growing in or by springs, for the habitat; *anti* (G) = against, plus *pyro* (G) = fire, referring to its use in Scandinavia to caulk chimneys.

Fontinalaceae

In a fast-moving brook, Sprague Pond Preserve, Phippsburg, ME

Description Stems ≤80 cm long are much branched and feathery looking with wide-spreading leaves, except at the slender tip where they are closely appressed. Leaves are 2–3 mm long, narrowly lanceolate, entire, ecostate, and concave, with inrolled margins making them appear narrow.

Similar Species Not likely to be confused with the distinctive *F. antipyretica*. *F. novae-angliae* is usually in slower-moving or still water, has much larger, wider leaves, and stem bases with leaves sparse or lacking. [*Fontinalis sphagnifolia*] shares the slender stem tip attribute illustrated below, but it has stem and branch leaves differentiated, versus much the same for this species, and it appears to be somewhat less common, though it may just be undercollected.

Range and Habitat Found from Newfoundland and Quebec south to Florida in moving water.

Name This species was first described in reference to a collection made in Dalecarlia, Sweden.

F. dalecarlica   *F. sphagnifolia*

Fontinalaceae

Top and right, under water, Sewall Pond, Arrowsic, ME; left, remains left high and dry in a seasonal streambed, Hancock, NH

Description Long, usually submerged stems ≤40 cm long (stems in the photo above are 10–15 cm), green to yellow-brown to copper; branches diverge from the main stem at roughly a 45-degree angle; leaves large, 2.5–5 mm long, ecostate, concave, and wide spreading, except appressed at tips, giving a pointed look to the branches as in *F. dalecarlica*, but with shorter tips. The wiry stem bases are apt to lack leaves in the older portions.

Similar Species Compared with the previous two species, these stems have bases more wiry, frequently tangled, and with long leafless sections. While the habitats will overlap, *F. novae-angliae* is usually found in more slowly moving (or still) water. *F. dalecarlica* has narrower, smaller leaves, and *F. antipyretica* has

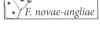

large, keeled leaves and branches with a very triangular cross section.

Range and Habitat Throughout the Northeast, from Newfoundland to Florida in slow-moving to still water.

Name The species was first described in reference to a collection made in Rockport, MA (New England).

F. novae-angliae

Fontinalaceae

Above (0.25×) and right (≈1×), on rock, Pope Farm, Nelson, NH; bottom right, capsule 5× by Bob Klips

Description Green with whitish overtones, irregularly branched, stringy when dry, with the base of the reddish stem generally lying against the substrate and rising up for the last 2 cm or so. As shown on the right, the moss changes appearance dramatically with hydration. The ecostate, clasping, concave leaves are appressed when dry and spreading when wet. The leaf tips are white because of a lack of chlorophyll. Capsules are short, stocky, and immersed in ciliate fringed leaves.

Similar Species Many of the rock mosses (*Grimmia* and *Schistidium* genera plus *Polytrichum piliferum*) have white leaf tips, but the white parts in those species are usually awns, or hair points, as opposed to this species, where the upper part of the leaf, not an extension, is white. Rock mosses frequently have white tips, presumably to reflect light and protect their chloroplasts from harsh sunlight. Many rock mosses, unlike this species, are costate acrocarps.

H. ciliata

Range and Habitat Throughout the Northern Hemisphere on exposed, noncalcareous rock, and on rooftops!

Name For Johannes Hedwig (1730–1799), whose posthumously published *Species Muscorum* is the starting point for moss nomenclature; *ciliata* is for the ciliate fringe of the perichaetial leaves.

Hedwigiaceae

Rich woods near Cherryfield, ME

Description A large, shiny, green to yellow-green moss in loose, occasionally dendroid mats with slender branches ranging from irregular to sparsely pinnate or bipinnate. The broad acuminate leaves are 1.5–2 mm long, have a double costa, are toothed and somewhat plicate when dry. The plications can look like a costa, as shown in the lower right photo. Stems are reddish and covered in paraphyllia that give a fuzzy appearance visible with a hand lens.

Similar Species Branching is not complanate or as neatly bipinnate as *Hylocomium splendens*. *Loeskeobryum brevirostre*, formerly *Hylocomium brevirostre*, is similar, but larger, with leaves less toothed, more acuminate, and widely divergent versus appressed to slightly divergent for this species. *Bryhnia novae-angliae* has a similar look but prefers wetter sites and is singly costate.

1 mm

H. umbratum

Range and Habitat From Labrador to Ontario, south to North Carolina; on humus over rock, or rotting wood in damp forests.

Name *umbra* (L) = shade, for the habitat.

Hylocomiaceae

Forest floor and tree base, Green Point Sanctuary, West Bath, ME

Description The splendid *Hylocomium* is beautifully bipinnate to tripinnate with large fronds 4–5 cm long. If you carefully untangle the fronds, you can see how each year's growth rises from the middle of the previous year's frond. Note the flattened branches and the red stem covered with matted paraphyllia. The plicate stem leaves have a short double costa (microscope required) and are 2–3 mm long with twisted tips. Primary branch leaves are 1–1.5 mm long, and secondary branch leaves are <1 mm long. The yellowish tint is typical. Capsules rare, mature in the fall.

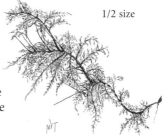

1/2 size

Similar Species *Thuidium delicatulum* is bipinnate with smaller fronds, costate leaves, green stems, and a dull look due to papillae, and it lacks this unusual growth form. *Loeskeobryum brevirostre* is singly pinnate with squarrose stem leaves. The occasional population of *H. splendens* lacking the "stair-step" look can be separated from *Pleurozium schreberi* by the paraphyllia along the stem, which *P. schreberi* lacks.

Range and Habitat Throughout the boreal forests of the Northern Hemisphere, south along the spine of the Appalachian Mountains to North Carolina; on soil and humus in shaded forests and ravines.

H. splendens

Name *hyle* (G) = wood, *kommotes* (G) = a beautifier; *splendens* (L) = shining or brilliant, thus a bright shining light of the forest. Common name: stair-step moss.

Hylocomiaceae

Soil over calcareous rock in a shaded, humid waterfall gorge, Ricketts Glen State Park, Benton, PA

Description A very large yellow-green moss with subpinnately branched, arched, red stems as long as 15 or 20 cm, leaves 2–3 mm long, with very small teeth (may look entire), long-acuminate, somewhat plicate, ecostate (short and double, not visible with a hand lens), stem leaves upright to divergent when dry, squarrose when wet, branch leaves upright to divergent when dry, widely divergent when wet; dioicous, capsules rarely produced.

Dry stem

Similar Species *Pleurozium schreberi*, another big pleurocarp with a red stem, has leaves much more appressed. *Hylocomium splendens*, a close relative, is bipinnate versus singly pinnate for *Loeskeobryum*. *Rhytidiadelphus squarrosus* is similar, but its growth form is upright versus more horizontal for this species, and the *R. squarrosus* branch leaves are as squarrose as the stem leaves, whereas the branch leaves in *Loeskeobryum* are divergent but not squarrose.

Hydrated stem

Range and Habitat Found throughout our range, but more common from Pennsylvania south, in rich, shady, high-humidity forests on soil, soil over rock, rotten wood, or tree bases.

L. brevirostre

Name Leopold Loeske (1865–1935) was a German bryologist who collected in Germany and the French and Swiss Alps; *bryon* (G) = a cryptogamic plant. *Brev* (L) = short, and *rostr* (L) = snout, for the short snout on the operculum of the rarely produced capsules.

Hylocomiaceae   Ca

North Traveler Mountain Trail, Baxter State Park, ME; drawings from Grout 1903

Description A very common, large, exuberant moss; red stems ≤10 cm, with bright golden tips. Branching is inconsistently pinnate, that is, messily pinnate. The golden leaves are ecostate (or with a microscopic double costa), broad, plicate, concave, imbricate, and tightly packed in a good example of julaceous. Capsules, infrequently produced, mature in fall.

2 mm

Similar Species *Loeskeobryum brevirostre* is also large, red stemmed, inconsistently pinnate, and with mildly plicate leaves, but its leaves are divergent, it has a more southern distribution, and it has a different habitat preference. *Hylocomium splendens* occasionally lacks the "stair-step" look and can be confusing; however, it has matted paraphyllia along the stem that are lacking in *P. schreberi*. See also *Pseudoscleropodium purim* for differences from that species.

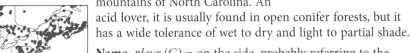

Range and Habitat From eastern Canada to the

P. schreberi

mountains of North Carolina. An acid lover, it is usually found in open conifer forests, but it has a wide tolerance of wet to dry and light to partial shade.

Name *pleur* (G) = on the side, probably referring to the pinnate branching; *schreberi* is for the German botanist Johann Christian von Schreber (1739–1810), a student of Linnaeus.

Hylocomiaceae

Above and right, Glidden shell midden, Damariscotta, ME; below right, found by Alison Dibble in a grassy spot behind the post office in Brooklin, ME

Description This large (stems ≤20 cm long), bright yellow-green, shaggy moss with its red stem, pinnate branching, upright form, and large (3.5–5 mm) plicate leaves with rugose tips crowded near the stem tips is tough to miss. The leaves have a double costa extending to the middle, though the plications make it hard to distinguish. Capsules are uncommon.

Similar Species *R. squarrosus* (*R. subpinnatus* of some authors), lower right, is a much less common but similar species with smooth, that is, not plicate, squarrose leaves. It seems to favor disturbed, grassy sites. *Loeskeobryum brevirostre* is closely related to and much like *R. squarrosus*, but it has branch leaves divergent and stem leaves squarrose, versus all squarrose for *R. squarrosus*, and it may be more common in the South. *L. brevirostre* has a more horizontal, tangled growth form than either of these species.

R. triquetris

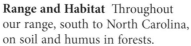
5 mm

Range and Habitat Throughout our range, south to North Carolina, on soil and humus in forests.

R. squarrosus

R. triquetris

Name *Rhytidium* is another moss genus, and *adelphos* (G) = brother, indicating a close relationship; *triquetrus* (L) = 3-angled, possibly for the shape of the pinnately branched plant.

Hylocomiaceae

Above and right, on limestone cliffs overlooking Lake Champlain, Lone Rock Point, Burlington, VT

Description A very large, bright, yellow-green moss in dense mats with main stems to 10 cm or more long, and, including leaves, 3–4 mm across; branching pinnate to irregular; leaves 3–4 mm long and, as the name implies, rugose, frequently falcate-secund at branch tips, as shown in the photo to the right, costate with costa 2/3 of the leaf length; never producing capsules in our range.

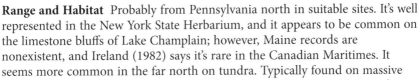

2 mm

Similar Species Looks much like a *Brachythecium*, but larger, with leaves falcate-secund, and without capsules.

Range and Habitat Probably from Pennsylvania north in suitable sites. It's well represented in the New York State Herbarium, and it appears to be common on the limestone bluffs of Lake Champlain; however, Maine records are nonexistent, and Ireland (1982) says it's rare in the Canadian Maritimes. It seems more common in the far north on tundra. Typically found on massive limestone outcrops and ledges in full to partial sun, often with *Abietinella abietina*. Sometimes on soil over the limestone rock.

R. rugosum

Name *rhytido* (G) = wrinkled, for the wrinkled leaves; *ruga* (L) = wrinkle, so the leaves are wrinkled in both Greek and Latin.

Hylocomiaceae   Ca

On rotten log, Green Point Preserve, Dresden, ME

Description A large, shiny, golden green moss with reddish-brown tinges; stems ≤8 cm, loose, mat-forming and stringy, to occasionally with upright branches, or very stringy and shaggy on tree bases. Irregularly pinnate with acuminate, ecostate (hard to see double costa), concave leaves 1.5–2 mm long. Leaves are loosely imbricate and not falcate-secund as in the *Hypnum* genus. Howard Crum (2004) said, "This species has short, tapered, somewhat flat branches resembling small swords." The frequently produced capsules are slightly constricted below the mouth when dry, with a 16-tooth peristome.

Similar species *Brachythecium* species are similar, and some will share the tree base and rotting log habitat, but they all have costate leaves and stocky capsules 2.5–3 mm long with red setae.

Range and Habitat Widespread from southern Ontario to the Canadian Maritimes, throughout northeastern North America, south in the mountains to Tennessee and North Carolina; on stumps, rotting logs, and tree bases. In the 1924 edition of his book, *Mosses with a Hand-Lens*, A. J. Grout said of *Callicladium haldanianum*, it "is almost sure to be found in any moist shady place where decaying wood is present, covering the unsightly masses of rotten wood with its upholstery of bright green."

Name *kalli* (G) = beauty and *klados* (G) = branch, thus beautiful branches; *haldanianum* is for D. Haldane, who collected the type specimen in Moose Factory, Canada, in 1825.

C. haldanianum

Hypnaceae

In water edge of limestone quarry, mixed with the
Calliergon giganteum shown on p. 222, Symmington
Quarry, Rockland, ME

Description Large plants in loose tufts,
shiny, green to yellow-green to brownish
green in older plants; stems ≤20 cm long,
irregularly pinnate with short side branches;
leaves wide spreading and somewhat
complanate on branches and appressed on
main stems, particularly at the cuspidate tip
(ending abruptly in a stout point). Leaves are
≈2 mm long, ecostate, concave to cucullate,
and end in a tiny sharp tip. Sporophytes are
rare.

Similar Species The *Calliergon* spp. in Amblystegiaceae are
similar but costate, and *Pleurozium schreberi* has a distinctive
red stem, versus yellow-green for this species, and is not
found in very wet habitats. See *Pseudoscleropodium purim*,
which also has apiculate leaves, but branch leaves much more
appressed, and a different habitat.

Range and Habitat Throughout our range, south to
North Carolina, in calcareous wet
meadows, swamps, and limestone
quarries.

C. cuspidata

Name *Calliergonella* refers to a small
(it's not) *Calliergon*, and *cuspidata* refers
to the shape of stem tips (see
description and illustration).

Along a gravelly roadside, Hancock County, ME; below, drawings from Grout 1903

Description Small, dark green to yellowish-green shiny moss with stems 1–3 cm long and squarrose, smooth, ecostate leaves (costa is microscopic—short, double, or absent completely) ≈1.5 mm long. Dry capsules are ridged and contracted below the mouth. The plants are monoicous, so they may produce capsules regularly.

Similar Species With similarly squarrose leaves, *Campyllium* spp. can be similar, but they have smooth capsules. In addition, *C. hispidulum*, which usually grows in wetter places, has leaves <1 mm. *C. stellatum* has a very different habitat. *Campyliadelphus chrysophyllus*, which has about the same size leaves and similar habitat, is costate. *H. striatella* was once placed in the *Plagiothecium* genus, but the current species included in that genus are more noticeably complanate.

Range and Habitat In the northern part of our range from southern Ontario to the Canadian Maritimes south to Michigan, and in the East, as far as Georgia at higher elevations; on moist, shady, noncalcareous cliffs, soil over rock, and clay banks.

H. striatella

Name *Herzogiella* is for Theodor Herzog (1880–1961), German bryologist, and *striatella* is for the small striate capsules.

Hypnaceae

Hypnum spp.

A. Stems with neat pinnate branching; sporophytes often present.................B
A. Stems irregularly branched; sporophytes rare...D

B. Plants small, stems and branches mostly <1 mm wide; stems green; sporophytes straight and upright, or nearly so; on bases of deciduous and coniferous trees, decaying wood, especially logs and stumps.. *H. pallescens*
B. Plants large, stems and branches often 1 mm wide or more; stems usually reddish; sporophytes curved or straight; frequently on rotten logs and stumps in woods, sometimes on humus, soil, and soil over boulders.......C

C. Plants with curved sporophytes...*H. curvifolium*
C. Plants with sporophytes straight and upright, or nearly so ......*H. imponens*

D. Branching moderately pinnate to less pinnate; leaves close and imbricate, concave, the stems and branches somewhat julaceous; on rock and trees, sometimes bases of trees, mainly in dry exposed places....*H. cupressiforme*
D. Few branches; leaves more distant, complanate to weakly concave, stems and branches not julaceous; on humus, and wet soil in roadside ditches, grassy meadows, beside lakes, and in swampy places................*H. lindbergii*

Hypnum imponens on soil over rock, with some *Dicranum* sp. poking through, Arrowsic, ME

Hypnaceae

Continued on next page

Hypnum spp.

Our common *Hypnum* species all share the characteristic of shiny, flat, falcate-secund, ecostate leaves with tips that wrap around the stem and point toward the substrate. Macroscopic characteristics that help separate them are branching pattern, stem color, size, capsules, and substrate. The most important microscopic characteristics are the shape and color of the alar cells.

Species	Branching	Stem color near tip	Primary substrate	Habitat	Capsules
H. cupressiforme	Partially pinnate*	Green	Calcareous rock	Open	Uncommon
H. cupressiforme var. *filiforme*	Partially pinnate*	Green	Non-calcareous rock and trees	Open to partial shade	Uncommon
H. curvifolium	Partially to very pinnate	Red-brown	Soil and rotting wood	Moist, shaded	Common, curved
H. imponens	Neatly pinnate	Red	Soil	Shaded forest floor, soil, soil over rock	Common, straight to slightly bent
H. lindbergii	Not pinnate	Green at tip	Soil	Open, grassy	Uncommon
H. pallescens	Partially pinnate	Pale green	Bark	Tree and tree base	Common, slightly curved

* *H. cupressiforme* can be quite neatly pinnate on rock, less so on trees.

Continued on next page

H. cupressiforme   *H. curvifolium*   *H. imponens*   *H. lindbergii*   *H. pallescens*

Hypnum pallescens, the tree base *Hypnum*, expanding its mission to include aboveground roots, Vinalhaven, ME

Hypnaceae

Above, *Hypnum cupressiforme* var. *filiforme* on noncalcareous rock; right, on cedar, Arrowsic, ME

Description This moderately pinnate *Hypnum* creeps over rock surfaces, its frequently parallel branches occasionally becoming so crowded that they are forced partially upright, causing it to lose its spreading appearance. On trees (right) the branches are narrow and stringy. The entire, ecostate, concave, shiny leaves are 1–2 mm long, appear tightly woven on the backs of the branches, and have long tips that curve toward the substrate, as is typical for this genus. The red-brown color component shown here seems typical.

Similar Species *Hypnum cupressiforme* grows on calcareous rock, whereas the var. *filiforme* shown here is found on noncalcareous rock and on tree trunks and branches. When on tree trunks, var. *filiforme* can easily be confused with *H. pallescens*, from which it differs primarily in microscopic characteristics, though *H. pallescens* is smaller and less tightly woven and has finely toothed flat leaves, whereas *H. cupressiflorme* leaves are entire and concave. See also range.

Range and Habitat *H. cupressiforme* seems restricted to dry calcareous rock, whereas var. *filiforme* is found on noncalcareous rock or tree trunks. Both *H.*

cupressiforme species are found in the Canadian Maritimes and on the coast of Maine. They are probably uncommon inland and south of the Maine coast.

Name *Hypnum* is an ancient Greek name for a moss thought to induce sleep or possibly was used for pillow stuffing, and *cupressiforme* is for a perceived resemblance to a cypress branch.

Hypnaceae   Ca in part

50 mm

2 mm

Above, on dead white birch, Pitston Farm, ME; below, from Grout 1903

Description Often in extensive clumps, pinnately branched, with prostrate reddish stems, and the usual shiny, acuminate, falcate-secund *Hypnum* leaves curled toward the substrate. The leaves are smooth, 2–2.5 mm long, loosely woven, and as the name indicates, they are very curved, often becoming circinate.

Similar Species The capsule (curved and ridged when dry) and microscopic characters (serrulate leaves, few enlarged alar cells and others) indicate that his species is most closely related to *H. lindbergii*, but that species has green stems, is not pinnate, and is more upright in form. *H. curvifolium* is more likely to be confused with *H. imponens* in the field, but it lacks the neatly tapered triangular shape of the *H. imponens* plants, and *H. imponens* capsules are straight.

Range and Habitat Throughout the Northeast near rivers and lakes, on moist soil banks, soil over rock, rotting wood, and trees.

H. curvifolium

Name *curvifolium* is for the very curved leaves, which occasionally become circular.

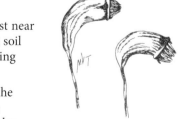

Furrowed capsules 2.5–3 mm long

Hypnaceae

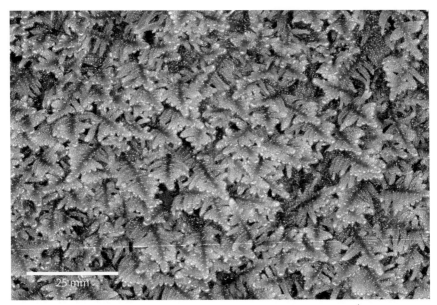

Pope Farm, Nelson, NH

Description This nicely pinnate *Hypnum* will cover large areas of a forest floor, creeping over rocks, logs and soil. The shiny, falcate-secund, ecostate leaves, 2 mm long, have very fine serrations in the upper half (hard to see), and turn their tips toward the substrate. The singly pinnate branches create a repeating pattern, as shown in the top photo. Capsules are erect and straight, maturing in winter.

Similar Species *Ptilium crista-castrensis* is similar, but with longer branches, a more vertical growth pattern, and pinnate branching continuing more completely to the tip, unlike our *H. imponens*, which seems to stop branching just before the tip, as shown in the hand lens photo to the right. *Sanionia uncinata* is less organized in its branching, and its leaves are costate and plicate. See also *H. curvifolium*.

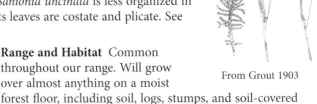

From Grout 1903

Range and Habitat Common throughout our range. Will grow over almost anything on a moist forest floor, including soil, logs, stumps, and soil-covered rock.

H. imponens

Name *Imponens* is from the same Latin root as imposter, for the fact that it can have many different looks. Common name: brocade moss.

Hypnaceae

On a gravelly, grassy, dirt road margin, Baileyville, ME

Description The bright green, frequently sparkly stems usually grow upward and are generally not pinnate at all. Probably because of the upright growth form, the branch leaves are less secund than in the other more prostrate *Hypnum* species. *H. lindbergii* leaf tips are curved toward a substrate but not necessarily well organized, and the leaves are less tightly braided on the side away from the tips. Leaves are shiny, 1–2.5 mm long. Stem is green at the tip and reddish brown below.

The least pinnate of our Hypnum species

Similar Species Capsules are ridged when dry, unlike all other *Hypnum* species (see the drawing of *H. imponens*) except *H. curvifolium*, but capsules are uncommon.

Range and Habitat Throughout our range, in wet habitats, frequently among grasses in damp meadows.

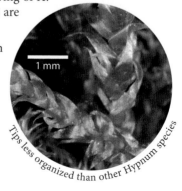

Tips less organized than other Hypnum species

Name *lindbergii* is for Sextus Otto Lindberg (1835–1889), a Swedish doctor and professor of botany.

H. lindbergii

Hypnaceae

H. pallescens

On base of a northern cedar, Arrowsic, ME

Capsules 1–2 mm long

Description Typically has a stringy look with some pinnate branching, though the branching is not as neat as in *H. imponens*. Light green stems ≤3 cm, with leaves <1 mm long, finely toothed, and as in all the *Hypnum* species, the leaves are shiny, ecostate, and curved toward the substrate. Capsules are common.

Similar species *H. imponens* and *H. cupressiforme* var. *filiforme* will also grow on tree bases, but both are larger with entire leaves and more regularly pinnate. *H. cupressiforme* var. *filiforme* is found only in the more northern part of our range, probably not much south of Maine. It also may not be as likely to have the stringy look on tree bases. *Pylaisiadelpha recurvans* is similar but somewhat complanate, with red stems.

Range and Habitat Very common on bases of hardwoods and cedar throughout our range.

Name *pallescens* may refer to the light green (pallid) stem color. Common name: tree-base moss.

Hypnaceae

On diseased beech in a rich hardwood forest, Millbridge, ME

Description Distinctively olive-green to yellow or brownish-green stems, with ascending branches, 1–2 cm long, with leaves ≈1 mm long, ecostate, acute to acuminate, slightly concave, entire, appressed when dry, less so when moist. Leaf tips are not falcate-secund. Stems are straight to somewhat curving dry or moist and often have microphyllous branchlets in leaf axils toward the tips.

Similar Species *Leucodon sciuroides* also grows on bark and has microphyllous branchlets but is larger and has plicate leaves ≈2 mm long. *Pylaisia* spp. are similar, with branches upturned, but they lack the microphyllous branchlets. *Leskeella nervosa* has similar branchlets, but is costate.

Note the microphyllous branchlets

2 mm

Range and Habitat Throughout our range, south to Florida on tree trunks and rotting wood.

P. repens

Name *platy* (G) = broad, *gyr* (G) = circle, for the broad annulus, the covering of the capsule; *repens* (L) = creeping, possibly for the primary stem that creeps over the substrate and gives rise to the more upright secondary stems.

Hypnaceae

25 mm

On vertical rock face, Squirrel Point, Arrowsic, ME

Description A bright shiny moss with loose, little-branched stems 2–3 cm long; leaves ecostate, complanate, wide spreading, entire, acuminate ≈2 mm long; filamentous, silky propagules formed in the leaf axils; dioicous with capsules probably not produced in eastern North America. Typically growing downward in sheets on a rock face.

Similar Species With the shiny complanate leaves and its seepy rock habitat preference, it looks and acts like a *Plagiothecium*, and indeed was once placed in that genus. The tiny vegetative propagules commonly formed in the leaf axils are the distinguishing characteristic.

Filamentous propagules

Range and Habitat Found throughout our range, south to the mountains of North Carolina on shaded acidic or calcareous rocks.

P. elegans

Name *pseudo* (G) = false, *Taxiphyllum* = a moss genus with complanate leaves (see *Taxiphyllum deplanatum*), for the similar look of that genus; *elegans* (L) = elegant or fine, for the soft silky look created by the fine propagules.

Hypnaceae

On a conifer forest floor on Mount Skatutakee Hancock, NH; lower right, Mary V. Thayer, from Grout 1903

Description Beautiful shiny green to yellow-green upright plants ≤10 cm tall, with neat, well-organized, pinnate branching, and falcate-secund, ecostate, long-pointed leaves pointing toward the branch below. The stem is green to yellow-green, and the pinnate branching continues completely to the tip.

Similar Species *Hypnum imponens* is smaller and more prostrate, and it has a reddish stem and pinnate branching that usually stops short of the tip.

Range and Habitat Throughout our range south to North Carolina, particularly in the damp conifer forests of the North Woods. On shaded humic soil and rotting wood.

P. crista-castrensis

Name *ptilon* (G) = a small feather, *crista* (L) = a crest, and *castrensis* (L) = military; thus a feather-shaped crest for a soldier. Common name: knight's plume.

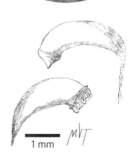

1 mm

Hypnaceae

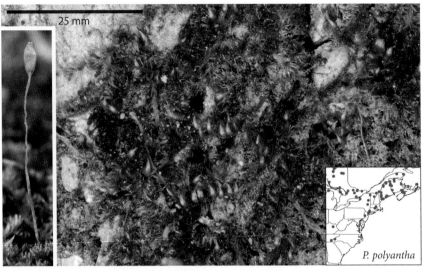

25 mm

P. polyantha

Above, on maple bark, Ravena, NY. Note how the branches flip outward, away from the tree trunk. Lower right, drawing by Mary V. Thayer, in Grout 1903.

Description A small, soft, yellow-green moss with stems to only ≈5 mm or so, ascending branches, and ecostate leaves ≈1 mm long or less, with long-acuminate, mostly straight tips. Monoicous, and a frequent producer of erect, straight capsules as shown, <2 mm long.

Similar Species Other pleurocarpous tree trunk species—*Leucodon sciuroides, Platygyrium repens,* and *Leskeella nervosa*—all have microphyllous branchlets (this does not), and the *Leucodon* is much larger. *Brachythecium* species are costate, and *Hypnum pallescens,* the most common tree bark *Hypnum,* is more organized, less tangled, and with leaves more falcate-secund. *Pylaisia polyantha,* [*P. selwynii*], and [*P. intricata*], closely related species with similar range and habitat, are best separated from this species by microscopic characteristics. That said, they are smaller, with branches more curled upward when dry. [*P. intricata*] leaves are falcate-secund and serrulate all around, while the other two are entire, or serrulate only at the tip. [*P. selwynii*] leaves are falcate-secund, and *P. polyantha* leaves are straight.

Range and Habitat Throughout our range, south to North Carolina, on deciduous tree trunks, reportedly with a particular fondness for sugar maple and poplar. Also on tree bases and rotten logs.

Name Bachelot de la Pylaie (1786–1856), a French explorer and scientist, was one of the first Europeans to collect in Newfoundland and on St. Pierre and Miquelon islands; *poly* (G) = many, *antha* (G) = flower, presumably for the many capsules produced.

Hypnaceae

Above, photo by Russ Kleinman and
Karen Blisard; right, leaves at ≈20× and
the stem at ≈10×, from Grout 1903

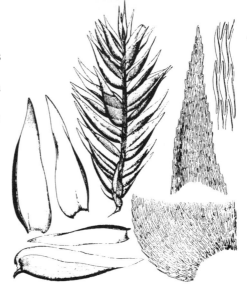

Description Medium-sized plants
with stems ≤4 cm long in thick to
thin yellow-green to yellow-brown
mats. Leaves are ecostate
(sometimes with short and double
costa), 1–2 mm long, ovate-
lanceolate, markedly complanate,
crowded, overlapping, somewhat
spreading but not squarrose.
Capsules are rare.

Similar Species This species
looks as though it belongs in
genus *Plagiothecium*, where it was
once placed. *Plagiothecium
cavifolium* also is somewhat
complanate and found on soil
over calcareous rock, but its leaves are more acuminate and have recurved tips.
The preference for such an uncommon substrate should separate this species
from most others. See also *Pseudotaxiphyllum elegans*.

T. deplanatum

Range and Habitat Apparently not common in our range
but possibly undercollected; on damp calcareous rock bluffs.

Name *taxi* (G) = arrangement, and *phyllum* (G) = a leaf,
combined with *deplanat* (L) = flattened to describe the leaf
arrangement.

Hypnaceae   Ca

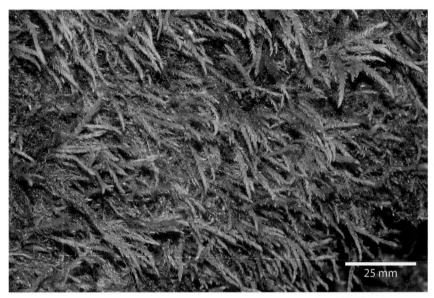

25 mm

Photos courtesy of Hermann Schachner; below right, Bry. Eur.

Description Prostrate stems 2–4 cm long with irregularly spaced, upright branches. Green to yellow-green with dark older stems. Stem leaves, ≈1 mm long, are appressed when dry, to erect-spreading when moist, gradually acuminate, and with a strong bulging costa that stops just short of the apex. Branch leaves are slightly smaller and less gradually tapered. The leaves are smooth to somewhat plicate and have a dull look when dry due to the presence of papillae. Capsules, 2–3 mm long, are commonly produced.

Similar Species The commonly produced capsules and the hardwood floodplain habitat help with the *Leskea* diagnosis. Other very similar species occur in the genus, but this species is the most common.

Range and Habitat Throughout our range, from the Canadian Maritimes south to North Carolina, on the bark of hardwood trees, often in floodplains.

L. polycarpa

Name *Leskea* is for Nathaniel Gottfried Leske (1751–1786), German naturalist and geologist; *poly* (G) = many, and *carp* (G) = fruit, for the many capsules.

2 mm

Leskeaceae

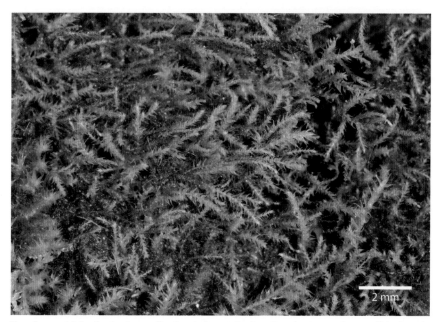

Top, from a Maine herbarium collection; below right, Bry. Eur.; bottom right, a well-hydrated plant from Hancock County, ME

Description A very small moss with branches <0.5 mm across, typically in pale to dark green tufts, with subpinnate branches rising from dark, underlying older stems. Stem leaves are ≈1 mm long, and branch leaves about half that. The strongly concave leaves are costate, erect-spreading when moist, to appressed when dry. Clumps of brood branches may be produced at the branch tips, though the small size makes them difficult to see. The capsules in the top photo are ≈2 mm long. Capsules are produced infrequently.

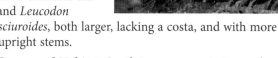

Similar Species Our other common mosses producing brood branches are *Platygyrium repens* and *Leucodon*

L. nervosa

sciuroides, both larger, lacking a costa, and with more upright stems.

Range and Habitat South in our range to Pennsylvania on tree bases and trunks, rotten logs, and occasionally on rock.

Name A diminutive of *Leskea* (previous page) and *nerv* (L) = nerve, for the strong costa.

Leskeaceae

In a high-humidity mature forest along a stream with many waterfalls, Ricketts Glen State Park, Benton, PA

Description Large yellow-green plants in loose tufts with stems growing out and upward in clumps; branching is subpinnate, with short, straight, side branches; leaves sometimes with a costa and sometimes not (!), ≤2 mm long, more or less plicate, acuminate, appressed to slightly spreading when wet; capsules with a hairy calyptra, reliably produced on setae that vary from very short with immersed capsules, to 1 cm long, as shown above. Unusual in having a variable costa, variable plications, and variable seta length; it is still quite identifiable by its robust form and commonly produced capsules.

Similar Species *Leucodon* spp. are very similar, and this species looks much like a robust *Leucodon*. *Forsstroemia* reliably produces capsules whereas *L. sciuroides*, the most common *Leucodon*, normally has none. In addition, the *Leucodon* calyptra is not hairy as it is for *Forsstroemia*, and *Forsstroemia* never produces the microphyllous branchlets found in *L. sciuroides*. *Platygyrium repens* is much smaller and has microphyllous branchlets, and *Pylaisia* spp. and *Leskeella nervosa* are much smaller. See also *Porella platyphylloidea*, a liverwort with something of a similar look and a tree-trunk habitat.

Range and Habitat From Massachusetts south to Florida and west to Louisiana in cool damp woods on bark of mature hardwoods, and occasionally on rock.

Name The genus name is for Johan Erik Forsström (1775–1824), minister and naturalist who collected widely in the Scandinavian Peninsula. He also finagled some warm weather duty, being assigned from 1802 to 1815 to St. Barts, now a French possession in the West Indies; *trich* (G) = hairs, and *mitra* (G) = a head dress, for the hairy calyptra.

Photos are from Cherryfield, ME

Description Brown-green to yellow-green, with usually round, unbranched, upright branches rising from prostrate main stems. Branches are ≤4 cm long with plicate, ecostate leaves ≈2 mm long, spreading when moist to more appressed when dry, as shown above. The branches curve out and upward when dry. The presence of clusters of microphyllous branchlets in the leaf axils, as shown above right, identifies *L. sciuroides,* the most common species in the northern part of our range. Capsules, partially immersed in surrounding leaves, are common in [*L. brachypus*], which is similar but lacks the microphyllous branchlets. Capsules are not known in *L. sciuroides.*

Similar Species [*L. julaceus*] lacks microphyllous branchlets, has smooth, not plicate leaves, is shiny, and has capsules held above surrounding leaves (see also range). *Platygyrium repens,* with very similar form, has microphyllous branchlets and shares similar habitat and substrate, but it is much smaller, with stems usually <2 cm long, branches with a less pronounced upward-hooking form, and smooth leaves; see also *Anomodon attenuatus. Pylaisia* spp. also have upturned branches but are much smaller and lack microphyllous branchlets. *Forsstroemia trichomitria* lacks microphyllous branchlets and reliably produces sporophytes. See also *Porella platyphylloidea,* a liverwort with a similar look.

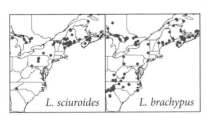

Leucodontaceae

Continued on next page

Range and Habitat [*L. brachypus*] ranges from the Canadian Maritimes south as far as Georgia and Alabama. *L. sciuroides* has a similar distribution but is uncommon south of Pennsylvania. [*L. julaceus*] is more southern in distribution, ranging from Pennsylvania south to Florida and Alabama. All species are found on deciduous tree trunks, in damp mature forests.

Name *leuco* (G) = white and *odon* (G) = tooth, for the light-colored peristome teeth common in this genus; *sciurus* (L) = squirrel, plus *oidea* (G) = looks like, for the bushy squirrel-tail look; *brachy* (G) = short, and *pus* (G) = foot, possibly for the short seta in *L. brachypus*. Common name for *L. sciuroides* is squirrel-tail Leucodon.

Leucodontaceae

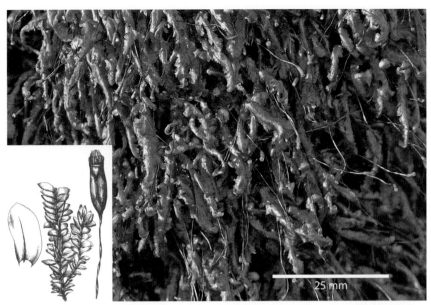

All photos by Hermann Schachner, courtesy of Wikimedia Commons; inset, from Grout 1924

Description A large moss with very complanate branches ≤5 cm long in shiny yellow-green mats; leaves 1–2 mm long, rounded, with a weak costa extending just past midway. Monoicous with capsules occasionally produced. The leaves look two-ranked, but they are actually inserted around the stem.

Capsules to 2 mm long

Similar Species With its complanate growth form and rounded leaves, it is easily confused with a liverwort (*Bazzania trilobata*), but no liverwort has a costa, and *Bazzania* has lobed leaves. The only other mosses that are this complanate and without pointed tips are *Fissidens* spp., which have complex folded leaves.

Range and Habitat From Newfoundland and Labrador south to North Carolina; on calcareous cliffs and boulders, and occasionally on tree (frequently cedar) bases.

Name *homos* (G) = alike or equal, and *alatus* (L) = winged, for the flat growth form; *trichomanoides* refers to the fern genus *Trichomanes*, which this species was thought to resemble.

H. tricho-manoides

On diseased beech in a rich hardwood forest, Millbridge, ME; bottom right, photo by Hermann Schachner, courtesy of Wikimedia Commons

Description These are large glossy plants that can vary from sparse to dense clumps or rings on tree trunks. The very complanate branches grow outward as much as 3–4 cm from the bark. Leaves are 2–3 mm long, shiny, and very noticeably wavy. Capsules are immersed and pendent from the underside of the stems, 1–2 mm long, on a very short seta ≈1 mm long.

Similar Species The flattened branches, wavy leaves, immersed capsules, and location relatively high (above tree base) on hardwood trunks make up a unique suite of characteristics. See also *Porella platyphylloidea*, a liverwort with something of a similar look (from a distance).

Range and Habitat Throughout our range south to North Carolina in cool, moist, deciduous forests,

N. pennata

typically on hardwood trunks at waist height and higher.

Name *Neckera* is for Noel Joseph de Necker (1730–1793), a bryologist contemporary of Johannes Hedwig; *penna* (L) = feather, for the pinnate arrangement of the leaves along the stems.

Neckeraceae

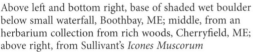

3 mm

Above left and bottom right, base of shaded wet boulder below small waterfall, Boothbay, ME; middle, from an herbarium collection from rich woods, Cherryfield, ME; above right, from Sullivant's *Icones Muscorum*

Description A large moss, dull green, somewhat inclined to upright. Its stems are dendroid, with lower stems devoid of leaves and with secondary stems 2–5 cm long with feathery pinnate branches frequently appearing blown to one side; leaves 1.5–3 mm long, toothed, costate (see below right), concave, and smooth. Branches are arranged in one plane, giving the plants a flattened aspect. Capsules are uncommon.

Similar Species *Climacium dendroides* is less one-sided in growth form, more likely to be upright versus inclined, and has plicate, shiny leaves versus smooth and dull for this species.

Range and Habitat Throughout our range south to Georgia. Probably less common than *Climacium dendroides* in the North, and more common in the South. On soil over rock in damp, shaded ravines.

Name *thamno* (G) = shrub, and *bryum* (G) = moss, for a shrubby moss; *alleghaniense* refers to the Allegheny Mountains, where this species probably reaches its maximum expression in hardwood coves.

T. alleghaniense

Neckeraceae

Squirrel Point, Arrowsic, ME

Description Glossy yellow-green plants with stems and branches either growing close to the substrate or ascending, as in this population. Also variable, the leaves can be crowded to distant, and complanate to less so. The leaves are reliably concave, entire, acuminate, symmetrical, and ecostate (short and double costa—see below right), they are 1.5–3 mm long, frequently with recurved tips. Overall branch width is usually >2 mm. Capsules are 1.5–3 mm long and not contracted below the mouth when dry, though this species is dioicous and capsules are not commonly produced.

Similar Species *P. denticulatum* and *P. laetum* are more consistently complanate, less likely to have ascending branches and thus more flattened against substrate, have leaves less concave, and are more tolerant of acidic habitats.

Range and Habitat Throughout the Northeast, south to Georgia in rich wooded sites on humic soil, soil over rock, and rotting wood.

P. cavifolium

Name *plagio* (G) = oblique, or slanting (also sometimes, sides or flanks) and *thec* (G) = a case, for the oblique capsule usually found in this genus; *cavea* (L) = hollow, for the concave leaves.

Plagiotheciaceae   Ca

Above and right, *P. laetum* along a little used trail, Arrowsic, ME. *P. denticulatum* has much the same look.

Description *P. laetum* and *P. denticulatum* are both glossy yellow-green plants with stems and branches growing close to the substrate. The leaves are entire, crowded, wide spreading, asymmetrical, flat, ecostate (double costa), and complanate, 1–1.5 mm long. Overall branch width is ≈ 2 mm. Monoicous with capsules 1.5–3 mm long. *P. laetum* dry capsules are smooth and not much contracted below the mouth, but dry *P. denticulatum* capsules (below) are wrinkled and contracted below the mouth. The fresh capsule typical of the genus has a yellow operculum with red tip. See next page.

Similar Species Unfortunately, the key characteristic separating these species is leaf cell size; *P. laetum* has a median leaf cell width of 5–7 microns versus 10–13 microns for the *P. denticulatum*. That said, *P. denticulatum* is darker, usually less glossy and less bright than *P. laetum*. *P. denticulatum* is often larger with slightly longer leaves, but much overlap is found in overall stem width and leaf length. See capsule differences discussed above. Fresh capsules are similar—see next page. *P. cavifolium* is more common in rich, limy sites with symmetrical, more concave, less complanate leaves. *Pseudotaxiphyllum elegans* has silky propagules in the leaf axils.

Range and Habitat Throughout our range to North Carolina, on mineral soil, and soil over seepy rocks and cliffs. Frequently along trails or seasonal rills where the plants are exposed by occasional water flow.

Plagiotheciaceae   Continued on next page

A mix of *P. laetum* and *Hypnum* spp. on the vertical, seepy rock face that is such a common site for both *P. laetum* and *P. denticulatum*. Fresh capsule shown is typical of the genus.

Name *laetus* (L) = bright and cheerful, for the bright, shiny appearance, and *denticulatum* for leaves toothed at the tip. The toothing is very minor and hard to pick up with a hand lens.

Plagiotheciaceae

Top, coniferous forest floor, Schoodic Peninsula, ME; below right, photo by Bob Klips, Blackwater Falls State Park, WV

Description A small plant with red to orange stems 2–5 cm long, complanate, with falcate-secund ecostate leaves and a *Hypnum*-type capsule. Branching is randomly pinnate, and the leaves are erect-spreading to appressed, with serrate tips and revolute lower margins. Crum and Anderson (1981) stated rather cavalierly that this species is "recognizable at a glance because of its extraordinary golden sheen, rivaled only by the green glossiness of *Plagiothecium laetum*." It appears silvery to others.

Similar Species This species looks very much like the *Hypnum* spp. to which it is closely related. The only *Hypnum* species with red stems are *H. imponens* and *H. curvifolium,* and both are much larger plants with larger leaves. *H. imponens* is neatly pinnate, and *H. curvifolium* is more or less pinnate. This species is more complanate than the *Hypnum* spp.

Range and Habitat A typical species of the spruce-fir zone, generally at higher elevations. Note, however, that these photos were taken within a few feet of sea level next to the ocean on the Schoodic Penninsula in Maine.

P. recurvans

Name *Pylaisia* for that genus and *delphys* (G) = womb, for the close relationship to the *Pylaisia* genus; *recurvans* is for the curved leaves. Formerly called *Brotherella recurvans*.

Sematophyllaceae

Above, from a Maine herbarium collection; color is unreliable. Drawing from Sullivant's *Icones Muscorum*.

Description A very small, light yellow-green to whitish green moss with stems <1 cm long, and leaves ≈0.5 mm, about as long as wide, ecostate, very concave, with an abrupt, toothed tip, and papillose cells (therefore very dull looking). The leaves are loosely erect, versus the tightly appressed leaves of its first cousin, [*M. julacea*]. Dioicous, but a frequent capsule producer.

Similar Species The less common [*M. julacea*] has a more julaceous (tightly packed) form, less of a leaf tip, and is not papillose, so it may be more shiny. The closely related and similar *Thelia hirtella* grows on tree trunks, is pinnately branched, is costate, and has ciliate (long-toothed) leaf margins.

Range and Habitat Throughout the Northeast and south to North Carolina in calcareous, frequently high-elevation rock crevices.

M. sibirica

Name *mys* (G) = mouse, *ur* (G) = tail, and *ella* (L) = diminutive, for the mouse-tail look of the moss; *sibirica* is for an early specimen found in Siberia. Common name: mousetail moss.

Above, on white oak bark, Squirrel Point Lighthouse Trail, Arrowsic, ME; right, from the NJ Pine Barrens

Description Small yellow-green moss with creeping primary stems and singly pinnate, densely packed, rising, julaceous secondary stems ≤6 mm long. Leaves are tiny (≈1 mm), costate with costa extending about halfway, broad, concave, short-pointed, and ciliate at the margin. The cells each have one large papilla, causing its dull, nonreflective appearance.

Similar Species With the tightly appressed, concave, ciliate-margined leaves, this species might be confused with a liverwort; however, liverworts never have a costa, and their leaves are two-ranked, with stems rarely looking as round as these.

Range and Habitat Throughout our range on hardwood bark, seeming to prefer white oak over all others. Rare in Québec and more common south. In Maine, this is a tree-base moss, but in the New Jersey Pine Barrens, where

T. hirtella

the two photos on the right were taken, it seems to reach its maximum expression and is common several feet above ground.

Name *thele* (G) = nipple, for the papilla on each cell, and *hirti* (L) = hairy, for the ciliate leaf margin. Common name: white oak moss.

Theliaceae

On soil over rock on a limestone cliff, Lake Champlain, Lone Rock Point, Burlington, VT

Description A large and showy moss, yellow-green to brownish, with wiry stems ≤10 cm long; branching is singly pinnate with branch leaves small, <1 mm long and closely appressed; stem leaves much larger, ≈2 mm long, plicate, costate, mildly divergent and with tomentum in the leaf axils. Papillae give the leaves a dull look and a brittle feel. Does not produce capsules in eastern North America.

Similar Species *Thuidium* species are 2–3-pinnate. The Hypnaceae family includes large, singly pinnate species, but they all have falcate-secund leaves. *Pleurozium schreberi* is less wiry, has ecostate leaves, and is less neatly pinnate. *Helodium blandowii*, *Calliergon* and *Calliergonella* spp. are all found in wet habitats. See also *Pseudoscleropodium purim*.

Range and Habitat Throughout northeastern North America on dry cliffs on soil over calcareous, frequently limestone, rock. Often found growing with *Rhytidium rugosum*. Reported as far south as Virginia. Western North American populations produce capsules.

A. abietina

Name For a perceived resemblance to the *Abies* genus of fir trees. Older publications refer to this species as *Thuidium abietinum*.

Thuidiaceae   Ca

H. blandowii

T. delicatulum

25 mm

Collected growing with *Thuidium delicatulum* in an abandoned gravel pit next to Augusta State Airport, Augusta, ME

Description A large, erect, green to reddish yellow, soft, singly pinnate pleurocarp with small (<1.5 mm), costate (costa extending along 4/5 of leaf), plicate, frequently pale leaves, and red–brown stems. Dense paraphyllia give the stem a cobwebby appearance that can be seen with a hand lens. Autoicous, and probably a good capsule producer, though this developing capsule (below) is the only photo I have. The mature capsule is ≈3 mm long, quite curved, constricted below the neck, and with 16 yellow teeth.

Note the dense paraphyllia, 10ˣ

Similar Species Resembles a large *Thuidium delicatulum* but is singly pinnate versus bi- to tripinnate. See above. The dense paraphyllia and the calcareous swamp habitat will separate this from other singly pinnate pleurocarps.

Range and Habitat Throughout much of North America, south to New York and Michigan in the East in calcareous swamps.

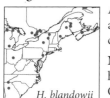

H. blandowii

Name *helos* (G) = marsh, for the habitat, and *blandowii* for Otto Christian Blandow (1778–1810), German botanist.

Above and right, wet woods in Nelson, NH; inset top left, leaf 10×, stem 20x, from Grout 1903

Description The branches of this beautiful, big, green to yellow-green moss are doubly pinnate, looking like miniature fern fronds. The dull (papillose), costate leaves are small (<1 mm), spreading when moist and imbricate when dry. Capsules are uncommon.

Similar Species [*T. recognitum*] is very similar, but is a calciphile, and its capsule operculum has a short snout versus a long one for *T. delicatulum* (capsules not illustrated). *Hylocomium splendens* is also doubly pinnate but is ecostate and has a unique growth form.

Range and Habitat Throughout North America on the forest floor in shaded, moist forests, particularly near streams or seeps. The substrate can be soil, humus, rotting logs, or soil over rock.

Name *thuja* (G) = cedar tree, for the flat sprays and regular branching of cedars, and *delicate* (L) = dainty, for the delicate twice-cut look. Common name: delicate fern moss.

Thuidiaceae

Scapania and associates. See page 347.

Liverworts and Hornworts

Liverworts and hornworts are generally less conspicuous than mosses and are covered here in less detail, with a focus on the more common genera, most of which are represented by an individual species.

The sizes given for each species and much other technical data are from Rudolf Schuster's monumental work, *The Hepaticae and Anthocerotae of North America*, published in 6 volumes between 1966 and 1992 by Columbia University Press (volumes 1–4) and the Field Museum of Natural History (volumes 5 and 6).

This illustration is one of 100 extraordinary plates in the 1904 book by German biologist Ernst Haeckel, *Kunstformen der Natur*. It illustrates European species, but many of them are also found in our range. Drawings 1–2, *Marchantia* spp.; 3–4, *Asterella* spp.; 8, *Lophozia* sp.; 9, *Cephalozia* sp.; 10, *Lepidozia reptans*; 13–15, *Scapania* spp.; 16, *Scapania nemorea*; 17, *Scapania* sp.

SPECIES INCLUDED

The order followed here, alphabetically by class or subclass within a simplified taxonomy, reflects the order of the species accounts. Species and genera in [brackets] are mentioned in keys, but not illustrated.

As used in this book, the term bryophyte includes mosses, hornworts, and liverworts. Current research is unclear whether division Bryophyta is a valid taxonomic entity, but it is a useful grouping, and most botanists continue to use it. Class Anthocerotopsida, the hornworts, is taxonomically separate from the liverworts and its species are treated here with them for convenience only.

Division Bryophyta
Subdivision Anthocerophytina
Class Anthocerotopsida, hornworts (p. 309)
   Notothylas orbicularis
   Phaeoceros carolinianus
Division Bryophyta
Subdivision Marchantiophytina
Class Jungermanniopsida
Subclass Metzgeriidae, simple thalloid liverworts (p. 311)
   [*Aneura pinguis*]
   Blasia pusilla (phylogenetic position uncertain)
   Metzgeria furcata
   Pallavicinia lyellii
   Pellia epiphylla
   Riccardia latifrons
Division Bryophyta
Subdivision Marchantiophytina
Class Marchantiopsida, complex thalloid liverworts (p. 316)
   Asterella tenella
   Conocephalum salebrosum
   Marchantia polymorpha
   Preissia quadrata
   Reboulia hemisphaerica
   Riccia fluitans
   Ricciocarpos natans
Division Bryophyta
Subdivision Marchantiophytina
Class Jungermanniopsida

Subclass Jungermanniidae, leafy liverworts (p. 324)
   [*Anastrophyllum*]
   Barbilophozia barbata
   Bazzania trilobata
   [*Blepharostoma*]
   Calypogeia muelleriana
   Cephalozia spp.
   [*Chiloscyphus*]
   Cladopodiella fluitans
   Diplophyllum apiculatum
   [*Fossombronia*]
   Frullania asagrayana
   Frullania eboracensis
   Geocalyx graveolens
   Gymnocolea inflata
   [*Harpanthus*]
   Jamesoniella autumnalis
   [*Jungermannia*]
   [*Lejeunia*]
   Lepidozia reptans
   Lophocolea heterophylla
   Lophozia incisa
   Marsupella emarginata
   Mylia anomala
   Nowellia curvifolia
   Odontoschisma denudatum
   Plagiochila porelloides
   Porella platyphylloidea
   Ptilidium ciliare
   Ptilidium pulcherrimum
   Radula complanata
   Scapania nemorea
   [*Tetralophozia*]
   Trichocolea tomentella
   [*Tritomaria*]

IDENTIFYING LIVERWORTS AND HORNWORTS

The good news is that liverworts, particularly at the genus level, have clearly defined characteristics, making them (usually) easier to identify than mosses. The bad news is that many liverworts are tiny! As with the mosses, size can be very subjective. For this section I have adopted Mary Lincoln's definitions as published in her excellent book, *Liverworts of New England*: sizes denote overall shoot width, including the stem and leaves.

Very small: <1 mm wide and <1 cm long; hard to see except under microscope.

Small: 1–2 mm wide and 1–2 cm long; hard to see except with a hand lens.

Medium-sized: 2–3 mm wide and 1.5–3 cm long; can be detected with naked eye.

Large: >3 mm wide and >3 cm long; easy to see.

This book does not attempt to provide guidance in identifying all liverworts. The concentration is largely on the more commonly found and more macroscopic genera. See the annotated references for in-depth liverwort guides.

The species are organized alphabetically within the 4 major categories: class Anthocerotopsida, the hornworts; subclass Metzgeriidae, the simple thalloid liverworts; class Marchantiopsida, the complex thalloid liverworts; and subclass Jungermanniidae, the leafy liverworts.

Throughout the species pages, [brackets] indicate a species that is not illustrated.

Key Characteristics to Note

Thalloid or leafy. The first big division in the liverwort world is thalloid versus leafy, and it's an easy one. Thalloid plants comprise a flat plate, or sheet, or ribbon of tissue lacking stems and leaves. Leafy plants have stems and leaves.

Sexual reproductive structures. Are the capsules simply held aloft by a translucent stalk or, as is true of some complex thalloid liverworts, are they held under a complex umbrella-like structure?

Simple capsules are characteristic of Metzgeriales, the simple thalloid liverworts, and Jungermanniales, the leafy liverworts.

Elaborate structures house the capsules and sometimes the antheridia in some of the Marchantiales, the complex thalloid liverworts.

What sort of structure protects the developing archegonia, and how are the antheridia housed? The spore-producing and disseminating capsules are, compared with mosses, ephemeral, but the protective sheaths, flaps, and sacs that house the developing capsules are more persistent. Note *Notothylas orbicularis*, *Pallavicinia lyellii*, and *Pellia epiphylla* for some examples.

Gemmae. Note presence or absence, location, and color of these frequently produced vegetative reproductive propagules.

Rhizoids. Location and description of these rootlike structures can be helpful for some species.

Color. Occasionally helpful. Most species develop more red-brown pigments in sun.

Habitat and substrate. Many species are quite habitat- or substrate-specific, or both.

Key Characteristics Specific to the Leafy Plants

Members of the order **Jungermanniales**, the leafy liverworts, account for most of our common liverwort species. The following characteristics are important in keying to genus or species:

Leaves. Unlike mosses, where leaves are arranged spirally around the stem, liverwort leaves are in two or three rows, giving a more flattened appearance. Look closely at the structure of the leaves. Do they consist of two connected parts, one on top of the other (complicate-bilobed), or are they simple? If complicate-bilobed, is the larger or the smaller lobe on top? If simple, is the margin of the leaf lobed, toothed, filamentous, or entire?

Complicate-bilobed small lobe on top   Filamentous   Lobed with underleaves and rhizoids, viewed from underside

Underleaves. Leafy liverworts frequently have a third row of smaller leaves on the underside of the stem. If present, note their size and shape. *Calypogeia muelleriana* underleaves are shown here.

Leaf arrangement. Do the leaves overlap in such a way that the upper margin of a leaf is under (succubous) or over (incubous) the bottom margin of the leaf above it? Or are the leaves inserted transversely so they are crosswise to the stem? Views below are from the tops of the shoots with the growth-tip pointed away from the viewer.

Incubous   Succubous   Transverse

Keying the Hornworts

Class Anthocerotopsida (hornworts) is a small group of small plants characterized by elongate spore-producing structures that split longitudinally when releasing their spores. They are infrequently encountered in our range, possibly because of their small size. I have included the two most common species here as representatives of this taxonomically important group. No further key is provided for this group—see the pages on hornworts in the species accounts.

DICHOTOMOUS KEY TO SPECIES OF THALLOID LIVERWORTS

1. Upper surface of thalli with pores and often with diamond-shaped markings..2
1. Upper surface of thalli with neither pores nor diamond-shaped markings...7

2. Thalli somewhat semicircular, upper surface with furrows; ventral surface with purple, lance-shaped, toothed scales; plants aquatic...*Ricciocarpos natans* (p. 323)
2. Thalli neither semicircular, nor with furrows; ventral surface lacking lance-shaped scales; plants not aquatic...3

3. Thalli small, usually <1 cm wide, margins usually purplish or reddish; diamond-shaped markings not visible with the naked eye; gemma cups lacking...4
3. Thalli large, often >1 cm wide, margins not purplish or reddish; diamond-shaped markings often visible with naked eye; gemma cups sometimes present on dorsal surface of thallus...6

4. Thalli with diamond-shaped markings clearly visible with hand lens; cells around the pores raised; margins plane, on calcareous substrate...*Preissia quadrata* (p. 320)
4. Thalli with diamond-shaped markings lacking or indistinct with hand lens; cells around the pores not noticeably raised...5

5. Thallus margins flat, female receptacles often present with dark red stalks.. *Asterella tenella* (p. 316)
5. Thallus margins raised or inrolled, female receptacles often present with stalks reddish to clear......................................*Reboulia hemisphaerica* (p. 321)

6. Thalli with gemma cups on dorsal surface; nonaromatic even when crushed; male and female receptacles often present, stalked...*Marchantia polymorpha* (p. 318)
6. Thalli lacking gemma cups; plants strongly aromatic, especially when crushed; male and female receptacles rarely present, the male sessile...*Conocephalum salebrosum* (p. 317)

7. Thalli with costa, and sometimes with small, dark spots from colonies of *Nostoc* (a cyanobacterium) within the thallus...8
7. Thalli lacking costa, or costa indistinct, without small, dark spots..............11

8. Thalli margins lobed...9
8. Thalli margins not lobed (sometimes gemmae on margins of *Metzgeria* look like small lobes) 10

9. Thalli with small dark spots (*Nostoc* colonies within the thallus) scattered along the margin, often in pairs; flask-shaped gemma receptacles often present near thalli apices.....................................*Blasia pusilla* (p. 311)
9. Thalli lacking dark spots and gemma receptacles.. *Pellia epiphylla* (in part) (p. 314)

10. Thalli large, ≥2 mm wide, lacking hairs on margin.. *Pallavicinia lyellii* (p. 313)
10. Thalli small, usually <2 mm wide, hairs on margin ....*Metzgeria* (p. 312)

11. Thalli thin and translucent, individual cells often evident, dichotomously branched.. 12
11. Thalli thick and individual cells not evident, irregularly branched............. 13

12. Thalli narrow, 1–2 mm wide; plants usually aquatic, submerged.. *Riccia fluitans* (p. 322)
12. Thalli broad, ≥3 mm in width; plants terrestrial.. *Pellia epiphylla* (in part) (p. 314)

13. Thalli sparingly branched, broad, ≥3 mm in width................ [*Aneura pinguis*]
13. Thalli much branched, narrow, <3 mm wide........................*Riccardia* (p. 315)

DICHOTOMOUS KEY TO GENERA OF LEAFY LIVERWORTS

1. Leaves divided into filaments or having margins with many long cilia..........2
1. Leaves neither filamentous nor with numerous cilia on margins, entire or broadly lobed, sometimes the lobes ending in a cilium or the leaf base with a few cilia 4

2. Plants small, stems <1 mm wide, irregularly branched, leaves divided to base into 3–4 filamentous lobes, on moist decaying logs, shaded rock faces, and sometimes the bases of trees..[*Blepharostoma*]
2. Plants large, stems usually >1 mm wide, pinnately branched, leaves divided into 2–5 lobes, the margins ciliate 3

3. Plants light yellowish green to whitish green, plumose, usually >1 cm wide, leaves with narrow lobes with branched cilia.................... *Trichocolea* (p. 348)
3. Plants green to reddish brown to purplish brown, usually not plumose, <1 cm wide, leaves with broad lobes with unbranched cilia............................
....*Ptilidium* (p. 344)

4. Leaves complicate-bilobed...5
4. Leaves undivided or lobed, but not complicate-bilobed 10

5. Leaves with the dorsal lobe smaller than the ventral lobe................................6
5. Leaves with the dorsal lobe larger than the ventral lobe..................................7

6. Leaf lobes narrowly elongate (lingulate), the ventral lobe 2–3× as long as wide .... *Diplophyllum* (p. 329)
6. Leaf lobes almost round to broadly ovate, the ventral lobe 1–2× as long as wide... *Scapania* (p. 347)

7. Stems without row of underleaves .... *Radula* (p. 346)
7. Stems with row of underleaves 8

8. Plants large, stems often >1 mm wide; underleaves undivided ....
....*Porella* (p. 343)
8. Plants small, stems mostly <1 mm wide; underleaves bilobed 9

9. Plants reddish brown or dark green; ventral lobes formed into a bucket or helmet shape...*Frullania* (pp. 330, 331)
9. Plants yellowish green; ventral leaf not bucket- or helmet-shaped ....
....[*Lejeunia*]

10. Underleaves present..11
10. Underleaves absent 22

11. Leaves incubous 12
11. Leaves succubous or transverse.. 14

12. Leaves entire or retuse at apex....................................*Calypogeia* (p. 326)
12. Leaves with 3–4 teeth or lobes.. 13

13. Plants large, stems 3–6 mm wide, flagellate branches present; leaves tridentate..*Bazzania* (p. 325)
13. Plants small, stems 1–2 mm wide, flagellate branches lacking; leaves 3–4-lobed into fingerlike segments..*Lepidozia* (p. 335)

14. Leaves entire or somewhat retuse at apex... 15
14. Leaves 2–4-lobed 16

15. Underleaves bilobed, often with a tooth on the base of the lobes...................
.... [*Chiloscyphus*]
15. Underleaves entire, lanceolate...*Mylia* (p. 339)

16. Leaves 3–4-lobed 17
16. Leaves 2- or, rarely, 3-lobed 18

17. Leaves with deeply channeled lobes, the margins broadly reflexed.................
.... [*Tetralophozia*]
17. Leaves without channeled lobes and reflexed margins......................................
.... *Barbilophozia* (in part) (p. 324)

18. Leaves distant, flat, the lobes obtuse or broadly rounded at apex; plants occurring in *Sphagnum* bogs or other acidic, subaquatic habitats ....
....*Cladopodiella* (p. 328)
18. Leaves mostly close, usually somewhat concave, the lobes acute to narrowly obtuse at apex; plants occurring mostly in drier habitats 19

19. Underleaves entire, attached on one side of base of lateral leaves....................
.... [*Harpanthus*]
19. Underleaves bilobed or ciliate, rarely entire, not attached to lateral leaves. 20

20. Underleaves 2-lobed, divided nearly to base, the margins without cilia ....
.... *Geocalyx* (p. 332)
20. Underleaves 2-lobed, divided 1/2–2/3 their length, usually ciliate, or underleaves entire.. 21

21. Rhizoids confined to bases of underleaves; underleaves ciliate........................
.... *Lophocolea* (p. 336)
21. Rhizoids scattered throughout ventral surface of stems; underleaves sometimes entire...*Lophozia* (in part) (p. 337)

22. Leaves undivided or nearly so 23
22. Leaves 2–4-lobed 26

23. Rhizoids purple or violet; leaves wavy or ruffled when dry ....
.... [*Fossombronia*]
23. Rhizoids not purple or violet, usually hyaline or brownish; leaves not wavy or ruffled when dry 24

24. Leaves usually with serrated margins....................................*Plagiochila* (p. 342)
24. Leaves with entire margins (female bracts sometimes ciliate) 25

25. Plants green to reddish brown; female bracts ciliate at base; perianth tapered to ciliate mouth; common rotten wood, soil, or rocks, never in streams.. *Jamesoniella* (p. 334)
25. Plants green, or some plants reddish with a border of enlarged cells; female bracts entire; perianth truncate or tapered, the mouth entire; infrequent, on rotten wood, soil, or rocks, sometimes in streams.................... [*Jungermannia*]

26. Leaves 3–4-lobed 27
26. Leaves 2-lobed.. 29

27. Leaf lobes all about the same size .... *Barbilophozia* (in part) (p. 324)
27. Leaf lobes unequal in size, the dorsal lobe much shorter than the ventral lobe.. 28

28. Leaf lobes entire, gemmae reddish brown, sometimes lacking ....
....[*Tritomaria*]
28. Leaf lobes toothed; gemmae green or yellowish green, usually present ....
....*Lophozia* (in part) (p. 337)

29. Leaves strongly concave and saclike, each lobe ending in a long slender cilium; occurring only on rotten wood....................................*Nowellia* (p. 340)
29. Leaves neither saclike nor ending in a long cilia; on wood and other substrates.. 30

30. Leaves transverse.. 31
30. Leaves succubous.. 32

31. Leaves deeply divided, 1/3–1/2 their length; gemmae usually present, reddish or purplish brown.. [*Anastrophyllum*]
31. Leaves shallowly divided, 1/8–1/3 their length; gemmae lacking.....................
....*Marsupella* (p. 338)

32. Leaf lobes obtuse to broadly rounded 33
32. Leaf lobes acute to acuminate.. 34

33. Leaves somewhat concave, about as broad as long, the lobes about equal in size.. *Gymnocolea* (p. 333)
33. Leaves flat, much longer than broad, the ventral lobe larger than the dorsal lobe.. *Cladopodiella* (p. 328)

34. Plants small, usually 0.5–1.0 mm wide, stems transparent; leaves deeply cleft, the lobes often touching at tips....................................*Cephalozia* (p. 327)
34. Plants large, mostly >1 mm wide, stems opaque; leaves shallowly cleft, without lobes touching at tips....................................*Lophozia* (in part) (p. 337)

QUICK LOOK AT THE METZGERIIDAE, SIMPLE THALLOID LIVERWORTS

Species in the **subclass Metzgeriidae** (simple thalloid liverworts) lack elevated parasol-like reproductive structures, and some species have a noticeable midrib. Cellular structure and pores are not generally visible with a hand lens. Spores are produced in globular structures on a seta (see right), as they are in the leafy liverworts.

Species	Size	Capsules	Gemmae	Midrib	Comments	Page number
Blasia pusilla	Medium	Yes	Yes	Faint	Flask-shaped gemma structures	311
Metzgeria furcata	Small	No	Yes	Yes	Distinctive marginal gemmae	312
Pallavicinia lyellii	Medium	Yes	No	Yes	Curved thallus branches	313
Pellia epiphylla	Large	Yes	No	Faint	Note involucral flap	314
Riccardia latifrons	Small	Yes	Uncommon	No	Knobby shoot calyptrae	315

QUICK LOOK AT THE MARCHANTIOPSIDA, COMPLEX THALLOID LIVERWORTS

Species in the **class Marchantiopsida** (complex thalloid liverworts) lack a midrib, and in some species individual polygon shapes and air pores can be seen, sometimes not even requiring a hand lens. Some have special parasol-like structures housing the sporophytes (example to the right is *Reboulia hemisphaerica*). The species in this group that do not produce these elaborate, elevated structures are aquatic, at least at times, and produce spores in hidden structures in air chambers in the thallus.

Species	Size	Raised reproductive structure	Gemmae	Aquatic	Comments	Page number
Asterella tenella	Medium	Female	No	No	White fringe hangs over capsules	316
Conocephalum salebrosum	Large	Rare	No	No	Large, visible cells, aromatic	317
Marchantia polymorpha	Large	Male and female	Yes	No	Gemma cups reliably present	318
Preissia quadrata	Large	Male and female	No	No	Obligate calciphile	320
Reboulia hemisphaerica	Medium	Female	No	No	Margins curl up into purple tubes	321
Riccia fluitans	Small	Immersed	No	Yes	Straplike	322
Ricciocarpos natans	Small	Immersed	No	Yes	Fan-shaped, purple rhizoid fringe	323

QUICK LOOK AT THE JUNGERMANNIIDAE, LEAFY LIVERWORTS (RANDOM ACCESS)

Random Access Key

A random access key enables you to narrow down the list of possibilities using the most obvious characteristics. The numbers following each category refer to the table on the following page and indicate the species that have that characteristic. To find your species, use the following key to narrow the search to two or three possibilities, and go to those pages.

For example, if your sample has entire, simple leaves, your choices are species 3, 11, 16, and 18. If you then notice gemmae, your choices are 3, 14, 18, 23, and 24. The species common to these two lists are species 3 (*Calypogeia muelleriana*) and species 18 (*Odontoschisma denudatum*). You can then refine the choice by looking at the insertion, or you can flip to the two pages in question and compare your sample with the photographs and detailed descriptions.

Size

A. Small: 4, 5, 7, 8, 9, 10, 12, 14, 17, 18, 22, 23
B. Medium: 3, 6, 7, 11, 13, 14, 15, 22, 24
C. Large: 1, 2, 16, 19, 20, 21, 24, 25

Leaves

D. Complicate-bilobed leaves, smaller lobe on top: 6, 24
E. Complicate-bilobed leaves with larger lobe on top: 7, 8, 20, 23
F. Filamentous: 21, 22, 25
G. Entire simple: 3, 11, 16, 18
H. Toothed: 19, 24
I. Lobed: 1, 2, 4, 5, 9, 10, 12, 13, 14, 15, 17

Insertion

J. Incubous: 2, 3, 6, 7, 8, 12, 20, 21, 22, 23, 24**
K. Transverse: 1, 4, 5, 14, 15, 17
L. Succubous: 4, 5, 9, 10, 11, 13, 16, 18, 19, 24**, 25

Substrate

M. Tree: 7, 8, 20, 22, 23
N. Rock: 1, 6, 10, 15, 19, 21, 23, 24
O. Soil: 2, 3, 6, 7, 18, 21, 25
P. *Sphagnum*/bog: 4, 14, 16
Q. Rotting wood: 9, 11, 12, 13, 17, 18, 22

Gemmae

R. Produces gemmae: 3, 14, 18, 23, 24
A blank cell in the gemmae column indicates that no gemmae are produced.

Continued on next page

Quick Look at the Jungermanniidae, Leafy Liverworts

	Species	Size	Leaf	Insertion	Substrate	Gemmae	Page number
1	*Barbilophozia barbata*	C	I	K	N		324
2	*Bazzania trilobata*	C	I	J	O		325
3	*Calypogeia muelleriana*	B	G	J	O	R	326
4	*Cephalozia* spp.	A	I	K, L	P		327
5	*Cladopodiella fluitans*	A	I	K, L	P		328
6	*Diplophyllum apiculatum*	B	D	J	N, O		329
7	*Frullania asagrayana*	A, B	E	J	M, O		330
8	*Frullania eboracensis*	A	E	J	M		331
9	*Geocalyx graveolens*	A	I	L	Q		332
10	*Gymnocolea inflata*	A	I	L	N		333
11	*Jamesoniella autumnalis*	B	G	L	Q		334
12	*Lepidozia reptans*	A	I	J	Q		335
13	*Lophocolea heterophylla*	B	I	L	Q		336
14	*Lophozia incisa*	A, B	I	K	P, Q	R	337
15	*Marsupella emarginata*	B	I	K	N		338
16	*Mylia anomala*	C	G	L	P		339
17	*Nowellia curvifolia*	A	I	K	Q		340
18	*Odontoschisma denudatum*	A	G	L	O, Q	R	341
19	*Plagiochila porelloides*	C	H	L	N		342
20	*Porella platyphylloidea*	C	E	J	M		343
21	*Ptilidium ciliare*	C	F	J	N, O		344
22	*Ptilidium pulcherrimum*	A, B	F	J	M, Q		344
23	*Radula complanata*	A	E	J	M, N	R	346
24	*Scapania nemorea**	B, C	D, H	J, K	N	R	347
25	*Trichocolea tomentella*	C	F	L	O		348

Note: The letters following the species names indicate characteristics in the random access key on the previous page.
* The smaller, upper leaf lobes of *Scapania nemorea* are incubous, but the larger, lower leaf lobes are succubous.

Liverwort and Hornwort Species Accounts

Presented alphabetically within the 4 major categories: class Anthocerotopsida, the hornworts; subclass Metzgeriidae, the simple thalloid liverworts; class Marchantiopsida, the complex thalloid liverworts; and subclass Jungermanniidae, the leafy liverworts

Collected by Jerry Oemig in Stow, MA, on soil in a fallow field

Description Light green, often translucent; with protruding and sometimes overlapping lobes in a rosette shape 2–7 mm across; capsules are produced in swollen outward-pointing sheaths; yellowish spores tint the tips of the capsules at maturity, which is reported from late September to late October in northeastern North America. The small size of this annual makes it an unlikely discovery before fall.

Similar Species Rudolph Schuster, hornwort and liverwort authority, offered the following in *The Hepaticae and Anthocerotae of North America*: "The problem with *Notothylas* usually is not in identifying it, but being able to find the plant: in most disturbed sites it tends to occur as single individuals and rarely is gregarious in small aggregations."

Range and Habitat Rare or very overlooked, might be found anywhere in eastern United States; often on disturbed sites on wet, fine-grained, mineral soil in late fall.

Name *notos* (G) = back, and *thyla* (G) = sack or pouch, for the sheaths (involucres); *orbicularis* is for the roundish growth form.

Note This tiny, almost ephemeral and possibly rare species is admittedly an unusual species to be included in a book such as this. It is included here to add a tiny bit of breadth to the hornworts, and to provide a search image to as wide a group as possible to see whether this species is indeed rare, or just overlooked.

class Anthocerotopsida, family Notothyladaceae

Along the margin of a little-used driveway in Arrowsic, ME; right, detail of involucres, and a *Blasia pusilla* lobe in the foreground

Description Thallus dark green, opaque, in uneven rosettes ≤3 cm across, with smooth surface and smooth margins; sporangia long, ≤3 cm tall, splitting along the length of the capsule, which becomes yellow-brown from yellow spores; involucres (basal sheath around the tall sporangia) are often constricted at the mouth; the male parts of the plants are in small pits on the thallus surface. The plants in North America are largely monoicous; that is, the male and female parts are on the same thallus. Dioicous plants that may be found are *Phaeoceros laevis* (many authorities make a subspecies distinction here, referring to *Phaeoceros laevis* subsp. *laevis* and *Phaeoceros laevis* subsp. *carolinianus*). Genetic studies indicate that this genus may include several cryptic species, that is, species that are separate genetically but not morphologically.

Similar Species [*Anthoceros punctatus*] also has a thallus ≤3 cm across and sporangia ≤5 cm tall, but it has highly divided margins, brown spores, a bumpy thallus, and flared involucre tops.

Range and Habitat Throughout the United States and most of southern Canada. Look for these plants late summer to fall when the sporangia (the "horns") have split and the yellow spores make them easier to see. Frequently found on damp, gravelly, disturbed soil, often with *Blasia pusilla*.

Name *phae* (G) = dark or dusky, *cer* (L) is from the Greek *keras* or horned, thus with dark-colored horns. The sporangia become brown after splitting and releasing the yellow spores. *Carolinianus* refers to North Carolina, where the type specimen was collected; *laevis* (L) = smooth, for the upper surface of the thallus.

class Anthocerotopsida, family Notothyladaceae

In a gravel driveway, Arrowsic, ME. Note the lobate margins, black *Nostoc* colonies, and flask-shaped gemmae-producing structures.

Description Thallus light yellowish green to green, reddish when in sun, branched, with branches clustered in a generally rosettelike pattern; medium-sized with segments 1.5–5 mm wide, with lobed margins and sometimes an indistinct whitish midrib; flask-shaped gemmae-producing structures commonly found near thallus margins; dark dots frequently paired at the base of lobes house colonies of *Nostoc* (a genus of cyanobacteria); dioicous, rarely producing sporophytes.

Similar Species A unique species in its own family, suborder, and, in some classifications, order.

Range and Habitat Northern temperate to low Arctic worldwide. In our coverage area, most commonly found north of Pennsylvania on moist loamy or gravelly banks or path margins.

Name The genus name is for Blasio Biagi, a botanically interested monk from Valombrosa, Italy, dates uncertain. The pre-Linnaean binomial was originally published in 1729 by Petro Antonio Micheli, who said (translated from the Latin), "This new plant we are pleased to call *Blasia* after Father D. Blasio Biagi, a monk of the order of Vallumbrosa, a solitary and industrious botanist who traveled our Tuscan pathways in search of indigenous plants"; *pusilla* (L) = small, which it isn't particularly, at least not in a liverwort frame of reference. Interestingly, this 1729 name has never changed.

subclass Metzgeriidae, family Blasiaceae

Photos by Bob Klips

Description Thallus small, yellow-green, straplike, 0.6–1 mm wide × 5–20 mm long, much branched, flat, and lying against the substrate, with midrib; marginal gemmae common (white arrow, right); hairs single, not in pairs, sometimes along the thallus margin and on the ventral surface (not evident in these photos); sporophytes not produced in North America.

Similar Species [*Metzgeria conjugata*] lacks the gemmae and has marginal hairs in pairs.

Range and Habitat From southern Ontario to the Canadian Maritimes, south to Louisiana; on deciduous tree trunks, occasionally on limestone cliffs and ledges; [*M. conjugata*] is found on similar substrates.

Name *Metzgeria* honors Johann Metzger (1771–1844), German copper engraver and art restorer from Staufen im Breisgau, Germany; *furcatus* (L) = forked, for the thallus branching pattern.

In a red maple swamp on mucky soil, Arrowsic, ME

Description Light translucent green with a conspicuous, lighter midrib, the medium-sized thallus consists of sparingly branched, wavy-margined, often curved ribbons 2–6 mm wide × 1.5–2.5 cm long; dioicous with archegonia housed in a fringed involucre (sheath) (right photo), and antheridia in fringed flaps along the midrib (lower right photo); gemmae not produced. Reported to have a fishy smell (though I cannot smell it).

Similar Species [*Moerkia hibernica*] is similar but rare, restricted to calcareous habitats, and more branched, it has a wavier thallus margin, and its antheridia are clustered toward the branch tips, not along the midrib.

Range and Habitat Throughout eastern North America from the Canadian Maritimes to Florida, extending west roughly to the Mississippi River; common in acidic swamps on mucky soil along watercourses.

Microscope Tip *P. lyellii* has an opaque central strand in the midrib, lacking in [*M. hibernica*].

Name The genus name commemorates Lazarus Pallavicini (1719–1785), an Italian cardinal. The specific name is for Charles Lyell (1767–1849), Scottish botanist and father of the now more famous Charles Lyell (1797–1875), geologist and influential friend of Charles Darwin.

subclass Metzgeriidae, family Pallaviciniaceae

On mucky soil in a waterfall spray zone, Ricketts Glen State Park, Bradford County, PA. Capsules produced on long setae, broken during collection.

Description Thallus light to dark green, reddish in sun, particularly along the midrib, opaque except at the slightly undulate margins, 8–12 mm wide × 3–5 cm long, with indistinct midrib; archegonia enclosed in a flaplike involucral sac (see arrow above); antheridia in bumps on the surface of the thallus; monoicous; rhizoids abundant.

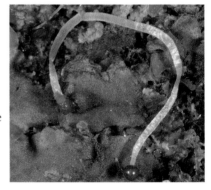

Similar Species [*Pellia neesiana*] is much the same, but dioicous, that is, the antheridia and archegonia are on separate plants. Most collections of [*P. neesiana*] are in northern New England and southeastern Canada, with very few collections farther south. It appears to be more common in western North America along the coast from southern California to Alaska.

Range and Habitat Widespread from the treeline in the North south to Georgia and Alabama; on mucky soil or wet rocks in shaded, acidic places.

Name *Pellia* commemorates Florentine lawyer Leopoldo Pellia-Fabroni; *epi* (G) = upon, and *phyllon* (G) = leaf, referring to the location of the reproductive organs.

subclass Metzgeriidae, family Pelliaceae

On rotting wood, Great Wass Island, Washington County, ME

Description Light to dark green thallus (color below right is a microscope artifact), small, much branched, with individual branches 0.5–1 mm wide, no midrib; monoicous, with capsules produced in knobby white structures (shoot calyptrae); gemmae uncommon.

Similar Species Other species of *Riccardia* are similar, and difficult to key out without more technical manuals.

Range and Habitat From southern Ontario to the Canadian Maritimes, south in eastern United States to New York and Pennsylvania. Other *Riccardia* spp. are more common to the south. Most common on rotting wood in cedar swamps and peat bogs.

Name The genus name is for Vincento Riccardi, eighteenth-century Florentine gentleman; *lati* (L) = wide, and *frons* (L) = leaf, for the branching pattern (lower right).

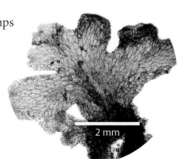

subclass Metzgeriidae, family Aneuraceae

5 mm

Photo by Bob Klips

Description Bright green to reddish with purple margins; medium-sized, with thallus straps 1.5–3.5 mm wide × 0.8–1.5 cm long, dichotomously branching, flat, and not curling upward when dry, no midrib; monoicous, with elevated female receptacles showing distinctive fringes on the capsules, male parts appearing as bumps on the thallus; no gemmae.

Similar Species Most other complex, thalloid liverwort species (Marchantiales) are larger. In addition, *Marchantia polymorpha* has elevated male structures and gemma cups, *Reboulia hemisphaerica* has elevated male structures and severely rolled-up margins when dry, *Preissia quadrata* is an obligate calciphile, has both male and female parts elevated on stalks, and has pores more visible with a hand lens. None of the similar species has the fringed female parts.

Range and Habitat Apparently restricted to the temperate forests of North America, and not found in the spruce-fir zone. Herbarium records show spotty distribution from southern Ontario east to southern New England, and south to Georgia and Louisiana; in the northern part of its range, found on damp soil in rock crevices, while reported more common on disturbed soil in the South. Desiccation tolerant.

Name *aster* (G) = a star, and *ella* (L) = small, for the elevated female structures; *tenellus* (L) = delicate, which I suppose it is.

class Marchantiopsida, family Aytoniaceae

On organic soil in a shady seep, Gulf Hagas, ME; below, from Grout 1924

Description Thallus flat, yellow-green to dark green, large, with branches 10–22 mm wide × 8–24 cm long, opaque, with poorly defined midrib; very leathery looking, with large macroscopic polygons separated by deep, dark reticulations, each cell topped by a single conspicuous pore; dioicous, with cone-shaped female receptacles (see drawing) rarely produced. Strongly aromatic when crushed.

Similar Species *Marchantia polymorpha, Preissia quadrata,* and *Reboulia hemisphaerica* all are much more likely to produce raised umbrella-like female receptacles, and none has such coarse cellular structure. *M. polymorpha,* found throughout the range of *C. salebrosum* in our coverage area, is very similar, differing in having a duller surface and less prominent airpores.

Range and Habitat Widespread throughout North America. Apparently favoring neutral to high pH sites in the northern part of the range, tolerating more acidic sites in the South. Found most often on damp soil along and in streams and seeps. In his 6-volume magnum opus on liverworts and hornworts, Rudolf Schuster described this species as widespread in ravines along springs and rills. Appropriately, the photo above was taken in Maine's most dramatic ravine.

Name *conus* (L) = a cone, plus *cephal* (L) = with a head, referring to the cone-head female structures rarely produced; *salebra* (L) = rough, and *osum* (L) = full of, possibly for the rough-textured look.

class Marchantiopsida, family Conocephalaceae

In a shaded garden pathway between bricks, Portland, ME. The elevated parasol-like structures are female receptacles, well past their prime. See next page for species description.

class Marchantiopsida, family Marchantiaceae

We get a little geographical diversity here. The shot above of an all-female colony was taken in northern England, as was the photo below of male receptacles mixed with a few developing females. The photos to the right and on the previous page were taken in Portland, ME.

Description Thallus large, flat, green to dark green, with individual straps 7–13 mm wide × 4–6 cm long; individual polygons each with 1 pore visible to the naked eye; midrib present but usually incomplete and spotted looking (see previous page); dioicous with frequently produced, very distinctive, elevated male and female receptacles on separate plants; gemma cups almost always present. See photos.

Similar Species *Conocephalum salebrosum* rarely produces elevated structures and has a strong smell and a leathery feel, unlike this species which has no smell and is not leathery. None of the other complex liverworts produces gemma cups like these.

Range and Habitat Throughout North America south of the Arctic; this very common species, is found in wet, disturbed, often burned sites. Frequently associated with enrichment of some type (e.g., lawn fertilizer or limestone).

Name Genus name is for Nicholas Marchant, seventeenth-century French botanist; *poly* (G) = many, and *morpha* (G) = forms, for the morphological diversity to be expected in such a widespread species.

class Marchantiopsida, family Marchantiaceae

Collected on shaded limestone processing waste in Rockport, ME; above, plant in situ; below, female receptacle; bottom, male receptacle

Description Thallus flat, dull green, large, ≤1 cm wide × 1.5 cm long, not branched or sometimes dichotomously branched, growing in frequently large colonies; cellular structure difficult to see without hand lens; midrib not noticeable; monoicous or dioicous, female receptacles (shown) elevated, and with 4 capsules, male receptacles also elevated, with flatter and more disklike shape, bumpy center, and hyaline margin; no gemmae produced.

Similar Species *Reboulia hemisphaerica* has raised thallus margins (sometimes curled up into a tube) and lacks raised male receptacles. This species never has the gemma cups common on *Marchantia polymorpha*.

Range and Habitat Throughout northern North America, south in our range to Pennsylvania; an obligate calciphile, most common on soil over calcareous rock, or on calcareous rocks in or near streams.

Name *Preissia* commemorates Balthasar Preiss (1765–1850), Austrian naturalist; *quadrata* is for the square look of the capsule from above.

class Marchantiopsida, family Marchantiaceae

Above and right, photos by Bob Klips; bottom right, photo by the author, showing the often dark purple margins

Description Thallus light green on top with dark margins, purplish below, ≤5 mm wide, with raised margins, curling up into a purplish tube when dry, midrib not noticeable; monoicous, female receptacles raised, antheridia (male parts) forming tiny, purplish, sausage-shaped structures around the base of the female structure; cell divisions and pores not visible with the naked eye; no gemma cups.

Similar Species *Preissia quadrata* is flatter, less dry sited, and an obligate calciphile, has both male and female parts elevated on stalks, and has pores more visible with a hand lens. *Marchantia polymorpha* has gemma cups, is less dry sited, is flatter, has both male and female receptacles elevated, and has much more visible cellular structure and pores.

Range and Habitat Throughout the United States and southern Canada; desiccation tolerant; on wet soil over rock, calcareous or not.

Name *Reboulia* commemorates Eugène de Reboul (1781–1851), French botanist, and *hemisphaerica* is for the round-topped female receptacle.

class Marchantiopsida, family Aytoniaceae

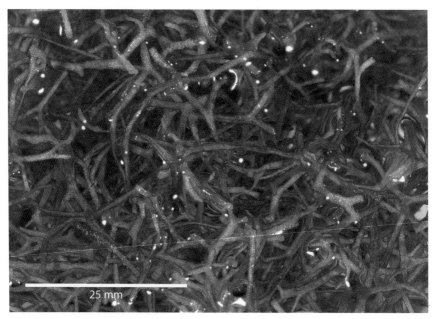

25 mm

This specimen is from Carolina Biological Supply—I was too impatient to wait until late summer for local plants!

Description Small, thallus comprising narrow, repeatedly dichotomously branched straps 0.7–2 mm wide, light to dark green, translucent, frequently forming tangled masses; reproduces by fragmentation; produces rhizoids only if stranded on soil. This is an annual in much of our range, growing from fragments of the previous year's plants. It is most frequently encountered in late summer or fall.

Similar Species Several other *Riccia* species occur in our area; all are small and generally not aquatic, though they are often found on muddy banks of still waters. *Ricciocarpos natans* (next page), also an aquatic species, floats on the water surface, has a ventral groove, and forms heart shapes or rosettes rather than the tangled mass of *R. fluitans*.

Range and Habitat Southern Canada and throughout most of the United States; in still ponds or quiet streams floating just below the water surface, or sometimes stranded as waters recede.

Name The genus name commemorates Pietro Francesco Ricci, a Florentine senator, and *fluitans* (L) = floating, for the habitat.

class Marchantiopsida, family Ricciaceae

Top and bottom, the terrestrial form stranded on muck; middle, the much smaller aquatic form. Collected by Jerry Oemig, in Stow, MA.

Description Small, aquatic forms ≈2 mm across and with a fringe of purple rhizoids; terrestrial segments larger, 5–10 mm × 4–8 mm without rhizoids and often coalescing into a fan shape or circular pattern. Both forms are bright green, with a groove down the middle of each segment; monoicous, with sex organs immersed in air chambers in the thallus; usually sterile; an annual in much of our range and more frequently encountered in late summer or fall.

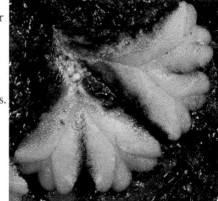

2 mm

Similar Species *Riccia* spp. lack the ventral groove and are more continuously branched, lacking the somewhat defined shape of this species.

Range and Habitat From southern Canada, south to Florida, Alabama, and Louisiana; floats on still waters, though often stranded; frequently growing with duckweed, a flowering plant.

Name *Riccia* is for that genus, and *carpos* is from *karpos* (G) = fruit, indicating that the spores are produced similarly, that is, internally in air chambers in these species; *natans* (L) = floating or swimming, for the habitat.

class Marchantiopsida, family Ricciaceae

On rock, Spruce Ledge, Temple, ME

Description Light to dark green, darker in sun, large, stems and leaves 2.0–5.0 mm wide × 3–8 cm long, little branched; with dense rhizoids on stem (see Name); dioicous; this species does not produce gemmae.

Leaves The squarish 4-toothed leaves are transversely inserted, though they often have a succubous look, as above.

Underleaves Small, inconspicuous.

Similar Species Other *Barbilophozia* species share the dense rhizoids, transverse to succubous insertion, and leaves with 2–5 lobes, and they frequently have reddish gemmae on leaf tips.

Range and Habitat Southern Ontario to the Canadian Maritime Provinces in the north, south in our range to North Carolina and Tennessee in the mountains; primarily on rocks and cliffs in conifer forests.

Name *barba* (L) = beard, for the dense rhizoids on the underside of the plant, and *lophozia* for a connection to that genus. The specific epithet reinforces the bearded look of the rhizoids.

subclass Jungermanniidae, family Scapaniaceae

Very common, the charismatic liverwort of the conifer forest, Bald Head, Arrowsic, ME

Description In dense mats with stems lying flat, as above, or often quite ascending; light to dark green; large, with shoots 3–6 mm wide, repeatedly forked and indeterminate in length; small, denuded (actually with some tiny leaves), flagellate branches common (see white arrows, right); few rhizoids; capsules uncommon. No gemmae.

Leaves Tightly overlapping and incubous, with 3–4 (predominantly 3) pointed lobes.

Underleaves Very small, slightly wider than the stem, with 3 (infrequently 4) pointed lobes, appearing to mimic the larger leaves.

Similar Species [*B. denudata*] is a bit smaller, has only 1 or 2 lobes, and has some portions of its stems without leaves.

Range and Habitat Throughout the forests of eastern North America, south to Florida; on humic soil or rotting wood in shaded conifer forests.

Name *Bazzania* commemorates Matteo Bazzani (1698–1749), professor of medicine at the University of Bologna, Italy, and supporter of botanical endeavors; *tri* (L) = three, plus *lobatus* (L) = lobed, for the leaves.

subclass Jungermanniidae, family Lepidoziaceae

On soil next to a root, near Goose Pond, Keene, NH; below, from Grout (1903) 1924

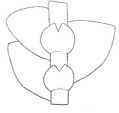

Description Light green to translucent blue-green; medium-sized with shoots 2.7–3.5 mm wide × 2–5 cm long, little branched; frequently ending in a cluster of gemmae, as above; our only genus with simple, entire, incubous leaves; monoicous, with elongate (cylindrical) capsules 3–4× as tall as wide.

Leaves Incubous, oval to slightly elongate, sometimes attenuate toward the stem tip.

Underleaves Notched, slightly larger than the stem (see right).

Similar Species [*C. integristipula*] is similar and found in similar habitats, but with underleaves less notched and >2× as wide as the stem, and with gemmae frequently found distributed along leaf margins, not tightly clustered as above.

Range and Habitat Appears to extend north to the treeline in Canada, south in our range to North Carolina and Tennessee in the mountains; commonly on humic soil.

Name *calyx* (L) = a flask, plus *hypo* (G) = beneath, and *geo* (G) = earth; these combine in the genus name in reference to a saclike structure (a marsupium) growing beneath the plant and housing the developing capsule. *Muelleriana* is for Karl Müller (1881–1955), German hepaticologist and author of *Die Lebermoose Europas* in 1951.

Note The name is occasionally spelled *Calypogeja*. Howard Crum (1981), noted bryologist and wit, wrote, "*Calypogeja* ... is orthographically defensible ... nomenclaturally correct ... and *Calypogeia* seems tedious." I took the more accepted, albeit tedious route.

subclass Jungermanniidae, family Calypogeiaceae

Cephalozia sp. growing with a *Calypogeia* sp. and *Scapania nemorea* (top left corner), near the Lead Mountain Trail, on soil along a streambank, Hancock County, ME

Description This is a genus of tiny plants with 2-lobed leaves. These species are most often seen among other bryophytes, frequently *Sphagnum* spp., under a microscope. They are probably impossible to identify to species in the field.

Leaves Leaves are <1 mm long with 2 sharp-tipped lobes, sometimes asymmetrical. Insertion is listed as succubous, but because of the spacing, it might be described as transverse or indeterminate.

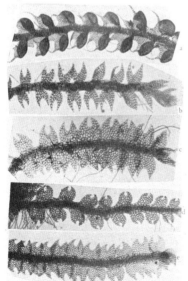

Underleaves None

Similar Species *Nowellia curvifolia* was once included in this genus (see below). *Cladopodiella* leaves are blunt-tipped.

Range and Habitat Throughout our range in very wet places, often in peat bogs.

Name *cephal* (G) = head, and *ozos* (G) = branch, describing the female branches, not illustrated.

Top to bottom: *Nowellia curvifolia* (once included in this genus), *C. bicuspidata*, *C. connivens*, *C. lunulifolia*, *C. catenulata*. Photomicrographs by Helen Greenwood, in Grout 1924.

subclass Jungermanniidae, family Cephaloziaceae

Collected with *Sphagnum* spp. in Molly Bog, Stowe, VT

Description Green to darker, tiny, overall shoot width <2 mm, tangled, little branched.

Leaves Leaves are <1 mm long with 2 round-tipped lobes, slightly asymmetrical. Insertion is listed as succubous, but because of the spacing, it might be described as transverse or indeterminate.

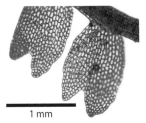

1 mm

Underleaves Tiny and irregular, but usually present near a stem tip.

Similar Species Very similar to *Cephalozia* spp., which have pointed lobes, versus rounded for this species.

Range and Habitat Commonly collected in *Sphagnum* dominated peat bogs throughout our range.

Name *klados* (G) = branch, and *podos* (G) = a foot, combined with the Latin diminutive *ella*, refers to a short branch at the base of the plant where sporophytes form; *fluitans* (L) = floating, for the bog habitat.

subclass Jungermanniidae, family Cephaloziaceae

On shaded rock, Sterling Gorge, Stowe, VT

Description Green in shade to reddish brown in sun; overall shoot width 1.5–4 mm, and ≤1.5 cm long; complanate.

Leaves Complicate-bilobed leaves with the smaller lobe on top in an incubous pattern, the larger lobe (ventral) diverging from the stem at roughly a 90-degree angle, with an abrupt tip (apiculus) on most large leaf lobes.

Underleaves No third row of underleaves.

Similar Species *Scapania* spp. also have the smaller lobe on top, but in *Scapania* the larger lobes are not as elongate.

Range and Habitat Endemic to eastern North America, from the Canadian Maritimes south to Georgia; found on rocks in streams, usually noncalcareous, and on gravelly, sandy, or clayey banks, not on humic soil or wood.

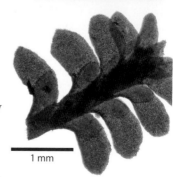

Name *diplo* (G) = two, and *phyll* (G) = leaf, for the folded-over leaves; *apic* (L) = apex, and *ula* (L) = little, for the abruptly pointed leaf tip.

subclass Jungermanniidae, family Scapaniaceae

On bark, Eagle Hill campus, Steuben, ME

Description Plants reddish brown, shiny, small to medium, shoots larger than *F. eboracensis*, 1–2 mm wide, indeterminate length, freely branched, often overlapping in dense mats.

Leaves Incubous, complicate-bilobed leaves, the larger lobe on top with a line of differentiated cells (ocelli) that can be seen with a 20× hand lens and good light (red arrow). As with *F. eboracensis*, the lower lobe is curved into a "bucket" shape (black arrow).

Underleaves Small, approximately 50 percent wider than the stem, with a distinct notch at the tip—can be obscured by the "buckets."

Similar Species Compared with *F. eboracensis*, this species is larger and more reddish, grows in denser, more overlapping colonies, grows on rock as well as on bark, and it has those sometimes difficult to see lines of differentiated cells. [*Jubula pennsylvanica*] shares the "bucket"-shaped underleaves but is found on wet, occasionally submerged rocks, is large, and has a more southern distribution, rare in New England.

Range and Habitat Dry sited and found on rock and bark throughout eastern North America from southern Canada to Florida, primarily in states that touch or are near the Atlantic Ocean. Note that *F. eboracensis* has a similar range with the addition of the states between this range and the Mississippi.

Name The species name memorializes Asa Gray (1810–1888), influential American botanist and professor of natural history at Harvard from 1842 to 1873.

subclass Jungermanniidae, family Frullaniaceae

On white birch, Nelson, NH; right, perianth, from Grout 1924

Description Plants dark green to brown, slender, small, stems including leaves are 0.5–0.9 mm wide × 1.5–3 cm long, irregularly pinnate and closely adhered to bark; dioicous, but in spite of separate male and female plants, frequently fertile and producing perianths, as shown right.

Leaves Incubous, large lobe on top with the lower lobe curled into a bucket shape (see previous page).

Underleaves Small, slightly wider than the stem, and with a slight notch at the tip; can be difficult to see under the "buckets."

Similar Species *F. asagrayana* (previous page) is larger, more pink, more densely packed on the bark, and with lines of reddish to clear cells on the leaves (seen with a 20× hand lens and young eyes). [*Jubula pennsylvanica*] shares the "bucket"-shaped underleaves but is large and found on wet, occasionally submerged rocks, and it has a more southern distribution, rare in New England. [*Lejeunia cavifolia*] shares the bark habitat and has complex-bilobed leaves, but is light green to yellow-green, is tiny, and has lower lobes not formed into buckets, often obscured by underleaves.

Range and Habitat Throughout eastern North America, east of the Mississippi River from southern Canada to Florida; on hardwood bark.

Name Named for Leonardi Frullani (1756–1824), Tuscan statesman; *eboracensis* (L) = of York, presumably referring to New York, where the type specimen was collected by Asa Gray, who is memorialized by *F. asagrayana*.

subclass Jungermanniidae, family Frullaniaceae

Above, a collection from Lisbon Falls, ME

Description Opaque, yellow-green with brownish stem; small, with shoots ≤1.5 mm wide × 2 cm long, unbranched, flat with leaves spreading horizontally; monoicous, but with sporophytes infrequent; no gemmae.

Leaves Succubous, flat, 0.8–1.4 mm long × 0.5–1 mm wide with 2 equal pointed lobes.

Underleaves Small with 2 long pointed lobes (see drawing, right)

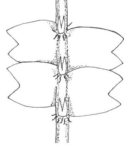

Similar Species *Lophozia* spp., also with sharp lobes and succubous insertion, usually have gemmae, and leaves that are curved or cupped but almost never flat. *Lophocolea heterophylla* is more translucent, with leaf tips more variable, and it regularly produces perianths.

Range and Habitat Throughout eastern Canada south of the treeline, south through our range following the Appalachian Mountains to North Carolina and Tennessee; on rotting wood and humic soil in wet, swampy places. Possibly favoring mildly calcareous sites.

Name *Geocalyx* is from *geo* (G) = earth, and *calyx* (L) = a flask, in reference to a rarely seen saclike structure (a marsupium) growing beneath the plant and housing the developing capsule (see also *Calypogeia*); *graveolens* is a common plant epithet for aromatic plants, from *gravis* (L) = heavy and *oleo* (L) = scent. Opinions of the smell of this plant vary from nasty turpentine to pleasantly aromatic. You be the judge.

subclass Jungermanniidae, family Geocalycaceae

On seepy granitic rock, Welch and Dickey Mountain, NH. *G. inflata* is the tangled, dark mass; lighter plants at the top of the photo are *Sphagnum pylaesii.*

Description Dark green to reddish black as above; small plants in dense tangled mats with shoots ≈1 mm wide with occasional branching; dioicous, but female plants often produce perianths; no gemmae.

Leaves The transversely inserted leaves are very cupped, almost globose, just >1 mm long, longer than wide, oval, with 2 rounded lobes. Some references list the insertion as succubous, but the leaves overlap infrequently, making this difficult to see.

Underleaves None.

Similar Species *Nowellia curvifolia* is about the same size and has similarly globose leaves, but its leaves have narrow filaments at the tips of the lobes, and it grows on rotting wood. [*Marsupella sphacelata*], an alpine species, has a similar ecology and, in sun forms, shares the dark brown-black color of *G. inflata.* [*M. sphacelata*] leaves also have 2 rounded lobes with a sharp sinus; however, [*M. sphacelata*] leaves are as wide as, or wider than, long, and may be more obviously transverse.

Range and Habitat Primarily a species of the boreal forest, the Arctic, and in high terrain throughout our range; often in a succession situation as above, a pioneer colonizing seepy, noncalcareous rock surfaces.

Name *gymnos* (G) = naked, *koleos* (G) = sheath, that is, the perianth is at the branch tip and not obscured by leaves (naked); *inflat* (L) = inflated, presumably for the inflated perianth, but a useful description of the balloon-shaped leaves.

subclass Jungermanniidae, family Scapaniaceae

Above, the green, clearly succubous plants: inset, typical reddish tint and upright growth at the stem tips, Tunk Mountain Trail, Hancock County, ME; below, drawing by Mary V. Thayer, in Grout 1924

Description Green to reddish green; medium-sized shoots 1.7–2.4 mm wide × 1–3 cm long, little branched; colorless rhizoids common; dioicous, perianths common, fruits in fall; no gemmae.

Leaves Succubous, leaves oval to round, ≈1 mm × 1 mm, divergent and clearly succubous along most of the stem, increasingly ascending and seeming to be transversely inserted as they fold upward near the tip.

Underleaves None.

One moist and two dry branches of *Jamsoniella autumnalis.*

Similar Species *Odontoschisma denudatum* has gemmae on partially denuded stems; *Mylia anomala* is restricted to peat bogs and usually has some gemmae on leaf tips; *Lophocolea heterophylla* has underleaves, is translucent, has 2 sharp-pointed lobes, and grows more closely adhered to its substrate.

Range and Habitat Throughout eastern North America, in Canada from Ontario to the Maritimes, south in the United States to Florida and Louisiana; on rotten wood in boggy areas.

Name The genus name commemorates William Jameson (1796–1873), Scottish physician and plant collector who lived in and collected widely in Ecuador; *autumnalis* refers to fall spore production.

subclass Jungermanniidae, family Jamesoniellaceae

Among tree roots, Eagle Hill campus, Steuben, ME

Description Light green to green in tangled, flat colonies; small, with shoots <1 mm wide; 1–2 pinnate branching.

Leaves Incubous, but the distinctive, deeply lobed leaves are sometimes not overlapping, rendering the incubous insertion a matter of some imagination.

Underleaves Much like the stem leaves, but less than half size.

Similar Species A. J. Grout (1924) said, "*L. reptans*, the common *Lepidozia*, is about the size of a *Ptilidium*, but is much less frequent and is in no danger of being confused with it, for the 3 to 4-cleft leaves curved downwards and looking like a half-closed hand are easily made out with a lens."

Range and Habitat Throughout eastern Canada south of the treeline, south through our range following the Appalachian Mountains to North Carolina and Tennessee; on rotting wood or humic soil.

Name *lepis* (G) = a scale, and *ozos* (G) = branch or twig, for the gnarly look of the female bracts (not illustrated); *reptans* (L) = crawling, for the low, creeping growth form.

subclass Jungermanniidae, family Lepidoziaceae

5 mm

Above, on rotten wood, Arrowsic, ME; below, photo by Hermann
Schachner, courtesy of Wikimedia Commons

Description Translucent green in shade (above) to
yellow-green in sun, growing closely appressed to its
substrate; medium-sized, with shoots ≤1.5 mm
wide × 1–3 cm long, some branching; rhizoids few, at
base of underleaves; monoicous, frequently producing
perianths (see photo right and Name); gemmae rare.

Leaves Succubous, wide spreading and flat to the
substrate, 0.75–1 mm long × 0.73–0.9 mm wide, bilobed
with lobe tips usually pointed, occasionally rounded, as
shown right, to sometimes unlobed.

0.5 mm

Underleaves Small, deeply lobed.

Similar Species *Mylia anomala* and *Odontoschisma denudatum* have rounded
leaves, produce gemmae reliably; *Jamesoniella autumnalis*, also with rounded
leaves, has no underleaves, is less appressed to its substrate, and is more opaque.
Geocalyx graveolens has leaves more opaque and more consistently sharp
toothed, though *L. heterophylla* is pretty consistently sharp toothed when on
rotten wood. *L. heterophylla* frequently produces perianths, whereas *G.
graveolens* does not produce them at all.

Range and Habitat Widely distributed and found throughout our range on
moist decaying wood or other organic substrates.

Name *lophos* (G) = crest, and *koleos* (G) = sheath, for the fringed perianth
mouth; *hetero* (G) = different, and *phyll* (G) = leaf, for the variable leaf tip shape.

subclass Jungermanniidae, family Lophocoleaceae

On soil, Great Wass Island Trail, Washington County, ME

Description Bright green to bluish green (black when dry in herbarium packets); shoots small to medium, 1.2–2 mm wide × 4–10 mm long, upright, in dense colonies, with an interesting "cabbage" sort of look; gemmae often formed on upper leaf margins; dioicous, with red capsules.

Leaves Transversely inserted, deeply incised leaves, 2–3 lobed, crowded and erect-spreading in heads (see right and bottom).

Underleaves None

Similar Species Several species of *Lophozia* occur in eastern North America; most have transverse leaf insertion and produce gemmae at the leaf tips, and all have lobed leaves. Some lab time and a desk reference will be necessary to sort them out.

Range and Habitat Southern Ontario to the Canadian Maritimes, south in higher terrain to Tennessee and North Carolina; on humic soil or rotting wood in wet woods or boggy places.

Name *lopho* (G) = crest, and *ozos* (G) = branch. Origin is unclear, but it works nicely for the crested shoot tips; *incisus* (L) = cut, for the incised leaves.

subclass Jungermanniidae, family Jungermanniaceae

On rock in a clear mountain stream crossing the Tunk Mountain Trail, Hancock County, ME

Description Green to reddish green or yellow-green; medium-sized shoots 1.8–2.4 mm wide × 2–5 cm long, upright from horizontal stolons; little branched, rhizoids only on stolons, no gemmae produced; dioicous.

Leaves Transversely inserted and diverging widely from the stem, 0.7–1.0 mm long × wide, with a shallowly rounded, bilobed margin.

Underleaves None.

Similar Species [*Marsupella sphacelata*] is similar but somewhat smaller, often quite black, and generally montane in distribution.

Range and Habitat From southern Ontario to the Canadian Maritimes, south in eastern North America to the mountains of Tennessee; on rock in or near mountain streams at all elevations.

Name *marsupium* (L) = a purse, for the puckered mouth of the perianth; *e* (L) = without, and *marginis* (L) = border or edge, referring to the indented or notched leaf apex.

subclass Jungermanniidae, family Gymnomitriaceae

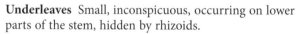

Mylia can be scattered among *Sphagnum* plants, or in a dense population. All photos were taken in a heath on Great Wass Island, ME.

Description Green in shade to yellow-brown (occasionally with reddish tints) in sun; large, with shoots ≤3 mm wide × 2–3 cm long, little branched; rhizoids abundant; gemmae frequently on the tips of the higher, more pointed leaves; dioicous.

Leaves Succubous. Round and wide-spreading lower on the stem, more erect and pointed higher on the stem. The succubous insertion becomes quite transverse as the leaves become more upright at the shoot tips.

Underleaves Small, inconspicuous, occurring on lower parts of the stem, hidden by rhizoids.

Similar Species *Jamesoniella autumnalis* has a very similar look but produces no gemmae and usually grows on wood. *Odontoschisma denudatum* produces tight clusters of gemmae on stem tips.

Range and Habitat In peat bogs with *Sphagnum* spp. throughout our range.

Name *Mylia* is for Willem Myliun, Dutch physician, and *anomalus* (G) = irregular, for the variation in leaf shape.

subclass Jungermanniidae, family Myliaceae

On rotting wood in the Yale Forest, Keene, NH

Description Green to reddish, very small with stems <1 mm wide, but frequently long, ≤2 or 3 cm long, unbranched; often completely covering a rotting log in shaded woods, giving it a tinted, felted look; dioicous, and often fruiting. Perianths shown on right.

Leaves Transversely inserted. As the species name indicates, the small leaves are very curved, resembling tiny Christmas tree balls to this observer. The leaves are 2 lobed, with each lobe ending in a slender tooth.

Underleaves None

Similar Species This species often shares its rotten log substrate with *Cephalozia* spp., which are also very small and 2-lobed but lack the inflated look.

Range and Habitat Southern Ontario to the Canadian Maritime provinces, south in our range to North Carolina and Tennessee in the mountains; a very common species always found on rotting, decorticate wood.

Name *Nowellia* commemorates the life of John Nowell (1802–1867), a largely uneducated millworker from Yorkshire, England, whose botanical skills brought him some measure of fame, if not fortune. Ironically, he did not collect liverworts. *Curvifolia* is, of course, for the very curved, almost globe-shaped leaves.

subclass Jungermanniidae, family Cephaloziaceae

On rotting wood in a treed swamp, Bald Head Island, Arrowsic, ME

Description Green to reddish green, small, shoots 1.0–1.6 mm wide × 1–2.5 cm long, little branched, with a distinctive reduction in leaf size toward the stem tip, which is covered with a cluster of gemmae; rhizoids on lower portion of stem; dioicous, sporophytes uncommon.

Leaves Succubous, round, cupped, often with a differentiated border, ≈1 mm diameter, smaller toward the stem tip.

Underleaves Small and inconspicuous, best found near the stem tip, but probably too small to see with a hand lens.

Similar Species Separated from the other entire-leaved, succubous plants with minimal or no underleaves (*Jamesoniella autumnalis* and *Mylia anomala*) by its apically denuded stem topped with a gemma cluster.

Range and Habitat Found sporadically from southern Ontario to the Canadian Maritimes, throughout eastern North America east of the Mississippi, north of Florida; on wet, rotting wood.

Name *odont* (G) = tooth, and *schism* (G) = splitting, for the perianth mouth, which has a ciliate opening, and *denudatus* (L) = made bare, for the somewhat barren stem tips.

subclass Jungermanniidae, family Cephaloziaceae

On a wet rock in a stream along the Tunk Mountain Trail, Hancock County, ME

Description Light to dark green, large but highly variable, with shoots 1.8–6 mm wide × 1–10 cm long, little branched, often upright; rhizoids occur on horizontal stolons but are absent from upright stems; no gemmae produced; dioicous.

Leaves Succubous, large, obovate, toothed (see detail photo), conspicuously rolled when dry and clearly extending down the stem (decurrency).

Underleaves None.

Similar Species This is our only succubous species with toothed, unlobed leaves. The rolled leaves and decurrencies are also distinctive.

Range and Habitat From southern Ontario to the Canadian Maritimes, south in the east to Pennsylvania, with an apparently disjunct population in the mountains of West Virginia and Tennessee; primarily on periodically inundated rock in shaded streams with no apparent pH preference.

Name *plagio* (G) = oblique, and *cheilos* (G) = edge, for the sometimes bent margin of the perianth. Howard Crum (1991) said of the species name that it means "uninformatively, or perhaps even inanely, *Porella*-like."

subclass Jungermanniidae, family Plagiochilaceae

On maple bark in rich woods near Cherryfield, ME

Description Yellowish to olive green, large, sparsely pinnate with branched sprays ≤4 cm wide × 10 cm long; hanging down and outward from tree trunks; leaves convex and tightly overlapping in an incubous arrangement.

Leaves Incubous. Folded over (complicate-bilobed) leaves with the larger lobes on top, with the smaller underside lobes looking much like the underleaves, giving a complex three-leaf look to the underside of the branches.

Underleaves With rounded tip, roughly the same width as the stem.

Similar Species *Frullania* species are smaller, more appressed to bark, and more likely to be reddish to dark. [*Jubula pennsylvanica*] is also large, has leaves that are complicate-bilobed, and has underleaves; however, it shares the characteristic of "bucket"-shaped underleaves with the *Frullania* spp. and is found on wet, occasionally submerged rocks and has a more southern distribution, rare in New England. See also the large pleurocarpous tree trunk mosses: *Neckera pennata*, *Forsstroemia trichomitria*, and *Leucodon sciuroides*.

Range and Habitat From southern Canada including the Maritimes, south in eastern United States to Florida and Louisiana; on tree trunks.

Name *Porella* is a misnomer for nonexistent pores in the capsules; *platy* (G) = flat or broad, *phyll* (G) = leaf, and *oid* (G) = like, that is, the plants look like [*P. platyphylla*], which has branches arrayed in broad, flat sprays.

Note This species has often been compared and confused with [*Porella platyphylla*], but recent studies have shown that [*P. platyphylla*] is not found in North America.

subclass Jungermanniidae, family Porellaceae

Description These, the two *Ptilidium* species in our range, share some characteristics and can be confusing. Both are pinnate with overlapping incubous leaves deeply divided into unequal lobes, with cilia on the margins. They can vary in color from light green to reddish brown to coppery red, and occasionally they can share substrate. Both species are dioicous, though only *P. pulcherrimum* produces sporophytes. *P. ciliare* is the larger of the two species; it grows upright frequently in deep tufts, and never (rarely?) develops sporophytes. Careful examination with a hand lens will give at least a glimpse of some undivided leaf lamina in this species. *P. pulcherrimum* is smaller, with more divided leaves, and more tightly appressed to its substrate, never ascending. Often with sporophytes, or at least with perianths (see photo below, right). The leaves of this species are so appressed and divided, it is usually impossible to find flat, undivided portions of leaf lamina with a hand lens.

Range and Habitat *P. ciliare* is found throughout Canada, south in eastern North America to Massachusetts, rarely farther south; *P. pulcherrimum* is also found throughout Canada but can be found south to North Carolina and Tennessee along the spine of the Appalachians. *P. ciliare* grows primarily on rock and soil over rock, only rarely on woody substrates, whereas *P. pulcherrimum* is usually on bark, wood, or some other organic substrate.

Name *ptilion* (G) = like a feather, combined with *-ium* (G), = small, referring to the usually twice-pinnate branching of these species; *cilium* (L) = hairs or hairlike, for the ciliate leaf fringe, and *pulcherimum* (G) = beautiful.

P. pulcherrimum capsule and perianth

Quick look comparing *Ptilidium* species

	P. ciliare	*P. pulcherrimum*
Size	Robust, stems ≤6 cm	Small–medium, stems ≤3 cm
Growth form	Upright, deep tufts	Flat
Leaf	Divided and ciliate, some lamina showing	Deeply divided, very ciliate with no lamina visible
Capsules	Rare	Common
Substrate	Rock, soil over rock	Bark, rotting wood, tree bases

subclass Jungermanniidae, family Ptilidiaceae

P. pulcherrimum, with highly divided leaves, tight to its woody substrate

P. ciliare, commonly found in deep colonies with upright stems on soil or soil over rock

subclass Jungermanniidae, family Ptilidiaceae

Spruce Ledge, Temple, ME, on rotting wood with dehisced capsules; middle, typical gemmae on leaf margins; bottom, the flattened perianth

Description Distinctive bright yellow-green plants, small, shoots 1.4–2.0 mm wide × 1–2 cm long, flattened, irregularly branched; monoicous and commonly with sporophytes, or at least perianths (bottom right), flattened, and with square tips; yellowish gemmae common on leaf margins (right).

Leaves Complicate-bilobed, incubous. The larger rounded lobe is on top; the lower lobe is squarish and about 1/4 the size of the top lobe.

Underleaves None

Similar Species Other conspicuous species with complicate-bilobed leaves with the larger lobe on top have underleaves and produce no marginal gemmae.

Range and Habitat Throughout eastern North America from southern Ontario to the Maritimes in Canada, and south to Georgia and Louisiana; on bark or rotting wood, occasionally on rock.

Name *radula* (L) = a scraper, for the shape of the perianth, and *complanatus* (L) = flattened, for the way the leaves and stems are flattened.

subclass Jungermanniidae, family Radulaceae

On a moist rock face, Eagle Hill campus, Steuben, ME. See also photo on page 292.

Description Shiny green to reddish brown, medium to large, shoots vary from 1.5 to 5.6 mm wide × 1 to 10 cm long, little branched; frequently with brown gemmae patches on leaves near the shoot apex (see small brown dots on plants in top photo); dioicous, rarely fruiting.

Leaves Complicate-bilobed, with the smaller lobe on top giving the appearance of 4 rows of leaves; margins toothed or spinose. The small upper leaf lobes are incubously arranged, while the large leaf lobes are succubous!

Underleaves None.

Similar Species *Diplophyllum apiculatum* has leaves more elongate, more wide spreading, more complanate, and with a short, sharp tip. Several other *Scapania* species will also key here.

Range and Habitat Throughout North America east of the Mississippi, from Ontario to the Canadian Maritimes, south to Florida and southwest to Louisiana; on seepy rock, rocks in streams or a spray zone, moist soil along watercourses.

Name *scapan* (G) = a shovel, for the shape of the perianth (not shown), and *nemoris* (G) = a woodland glade or pasture—not a good habitat descriptor.

subclass Jungermanniidae, family Scapaniaceae

Rich woods on humic soil, Mossy Glen Falls Trail, Stowe, VT

Description Large plants, light yellow-green to whitish green when dry, 5–10 cm long × 1.5–3 cm wide, double to triple pinnate; very woolly looking because of the highly divided, ciliate-margined leaves, which are succubous, though the structure of the leaves makes the insertion difficult to discern; dioicous, usually without sporophytes.

Similar Species *Ptilidium ciliare* is similar, but this species is larger with an even more tomentose look, lighter in color, less likely to be reddish, less upright, and restricted to high-pH environments.

Range and Habitat In eastern North America from southern Canada south to Missouri, Tennessee, and North Carolina; In cedar swamps or other damp, rich (high-pH) woods, on the forest floor.

Name *Trichocolea* is from *thrix* (G) = hair, and *koleos* (G) = sheath, for the covering of an emerging sporophyte (not illustrated and not commonly produced); *tomentella* is from *tomentum* (L) = woolly for the felted look of the plant.

subclass Jungermanniidae, family Trichocoleaceae

Illustrated Glossary

acrocarpous—Plants in cushions or tufts, main stems erect, or nearly so, simple or occasionally with forked branching, rarely **pinnate**, and if so, with short tuftlike branches; **costae** prominent and single; with **sporophytes** rising from stem tips.

acumen—A narrow, tapering leaf tip.

acuminate—Refers to narrow, tapering leaf with a concave margin.

acute—Sharp-pointed, but less so than **acuminate**.

alar cells—These cells at the corners of the base of the leaf are often differentiated, frequently quadrate or inflated appearing, occasionally colored. Viewing details of the alar region requires a compound microscope, but the presence or absence of differentiated alar regions can often be determined with a hand lens or dissecting microscope if the leaves are sandwiched between two microscope slides. See page 17.

antheridium, pl. antheridia—Male reproductive organ.

apical—At the tip.

apiculus—An abrupt point on a leaf tip.

apiculate—Having an **apiculus**.

appressed—Lying tightly against a stem.

archegonium, pl. archegonia—Female reproductive organ.

attenuate—Tapering, as in a branch getting more narrow toward the tip.

autoicous—Male and female organs are on the same plant, but in separate inflorescences. A form of **monoicous**.

awn—Filamentous leaf tip usually formed by an extension of the **costa**.

axil—The angle where a leaf attaches to a stem.

axillary—In the angle between leaf and stem, that is, in a leaf **axil**.

bog—Frequently a general term for a very wet place, but used here to mean an **ombrotrophic** bog; that is, a bog with minimal groundwater flow, getting its nutrients primarily from rain, snow, or other atmospheric deposition.

brood branches—Small branches functioning as vegetative **propagules**.

bryophytes—The group of plants comprising mosses, liverworts, and hornworts.

bulbil—A small vegetative **propagule**.

bulbiform—In the shape of a plant bulb. See *Funaria hygrometrica* basal leaves.

calciphile—Favoring calcium-enriched habitats, that is, with neutral to higher pH.

calyptra—Covering of a developing **capsule**, created from **archegonium** tissue. These structures take many forms and are often important in identification. See illustration at **capsule**.

capitulum—Upper section of a *Sphagnum* plant, often with densely clustered branches.

capsule—An often elevated structure in bryophytes where spores are made and distributed.

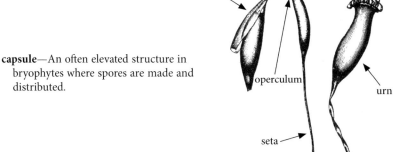

cladocarpous—With **sporophytes**, rising from the tip of a branch. Compare with **acrocarpous** and **pleurocarpous**.

columella—The central axis of a **capsule**.

complanate—Appearing flattened.

complicate-bilobed—Refers to liverwort leaves comprising 2 lobes folded together. In some species, one lobe may form into a saclike structure.

contorted—Twisted or curved, or both. See *Atrichum crispum* species page for comparison with **crisped**.

costa, pl. costae—Moss leaf midrib.

crisped—Strongly twisted or curved, or both. Compare with **contorted**.

cucullate—Hooded.

cuspidate—Refers to moss leaves ending in a short **apiculus**, or to pointed branch tips.

decurrent—Where leaf margins, or **costae**, extend down a stem past the point of leaf insertion, often creating a ridge on the stem.

dehisce—Rupture.

dendroid—Treelike, with branches all around the stem (see *Climacium* spp.).

dioicous—With male and female structures on separate plants.

diploid—With 2 sets of chromosomes.

distichous—Refers to leaves in 2 opposite rows along a stem.

divergent—Refers to leaves that do not lie close to stem, that is, they diverge from the stem.

divergent-pendent—Refers to *Sphagnum* stem leaves that grow downward from their attachment point and diverge from the stem.

ecostate—Lacking a **costa**, or midrib.

entire—Lacking teeth or indentations.

erect-appressed—Refers to *Sphagnum* stem leaves that grow upward from their attachment point and are close to the stem.

excurrent—Refers to a **costa** that extends beyond the tip of a moss leaf, creating some sort of extension or point. See **apiculus**.

exserted—Held above surrounding leaves.

falcate-secund—Refers to leaves that are curved (falcate) and aligned (secund). See Dicranaceae for examples.

fascicle—Refers to clusters of branches in Sphagnaceae.

fen—Wetland with some groundwater flow. Compare with **bog**.

filiform—Threadlike, slender.

flagellum, pl. flagella—Thin whiplike branches with or without vegetative **propagules** attached. *Aulacomnium* spp. and *Bazzania trilobata*; **flagelliform, flagellate**.

flexuose—Wavy.

frondose—With regular and flattened branching, as opposed to **dendroid,** with branches all around the stem.

gametophyte—The part of the plant that produces the sex organs (gametangia). In bryophytes, this is the green elaborative phase of the plant.

gemma, pl. gemmae—Vegetative **propagules** capable of producing a new plant; can take many forms, from simple buds to more complex, leaflike structures.

gemmiferous—Bearing **gemmae**.

globose—Spherical.

haploid—With one set of chromosomes.

homoiohydry—Capacity to maintain interior water content relatively independent of external variation in moisture levels; contrast with **poikilohydry.**

hyaline—Clear or translucent, colorless.

imbricate—Refers to leaves held closely against the stem and overlapping. See **julaceous**.

immersed—Refers to **capsules**, held on a **seta** so short that the capsule does not extend above the surrounding leaves (**perichaetial** leaves).

incubous—In a dorsal view, with leaves overlapping such that a leaf lies on top of the next leaf toward the tip of a branch.

involucre—In hornworts and **thalloid** liverworts, thalloid tissue surrounding and protecting an **archegonium**.

julaceous—Cylindrical, with closely **imbricate**, leaves, ropelike.

keeled—Sharply folded along a midrib.

lacerate—Appearing torn.

lamella, pl. lamellae—Longitudinal rows of stacked cells on the dorsal surface of some moss leaves. See Polytrichaceae.

lamina, pl. laminae—The flat part of a leaf blade, not including the **costa**.

lanceolate—Shaped like a lance, with straight sides tapering gradually to a point. See illustration at **alar cells**.

lignin—Woody material lacking in bryophytes.

lingulate—Strap- or tongue-shaped.

microphyllous—With very tiny leaves; see **brood branches**.

mitrate—Shaped like a mitre, in reference to a moss **calyptra** that is symmetrical and with a cone-shaped top.

monoicous—Male and female organs on the same plant; contrast with **dioicous**.

mucronate—Refers to a leaf with a small bump (mucro) at the tip, that is, a short **apiculus**.

muticous—In reference to the Grimmiaceae (rock moss) family, refers to the characteristic of leaves lacking a white (clear) leaf tip where one would normally be expected.

oblique—Angled; when applied to a leaf base, it indicates one side of the base is shorter than the other, that is, asymmetrical.

obovate—Egg-shaped, but with the wider part at the top.

obtuse—Blunt.

ombrotrophic—Without groundwater flow, that is, with all nutrients from above (from *ombros* (G) = rain, and *trophon* (G) = food); see **bog**.

operculum—A lid covering the **capsule** opening; see illustration at capsule.

ovate—Egg-shaped in two dimensions

ovoid—Egg-shaped in three dimensions.

papilla, pl. papillae—Bumps or more complex projections on the surface cell walls that scatter light and give leaves a dull appearance.

paraphyllium, pl. paraphyllia—Small **filiform** to leaflike structures on moss or liverwort stems or branches.

pellucid—Translucent to transparent.

percurrent—Refers to a **costa** reaching to a leaf tip but not beyond.

perianth—Sheathlike structure formed from leaf tissue that protects the **archegonia** and developing **sporophyte** in liverworts.

perichaetial—Refers to the often differentiated leaves surrounding the **archegonia** that give rise to a **seta** (almost nonexistent here) and **capsule**.

perichaetium—Perichaetial leaves plus the **archegonium**.

peristomate—Having a **peristome**.

peristome—Ring of teeth around the mouth of a moss **capsule** that assist in spore dispersal; see illustration at **capsule**.

pinnate—With branches frequently and evenly distributed on opposite sides of a stem; feather compound.

plane—Flat.

pleurocarpous—Plants in mats, the main stem prostrate or nearly so, sometimes with erect tips or branches, usually much branched, often **pinnate**, rarely simple, and if so, the stems long and intertwined; **costae** present or lacking; **sporophytes** rise from the tip of a very reduced and inconspicuous branch along the main stem somewhere below apex of a major branch.

plicate—With longitudinal pleats or folds.

plumose—Neatly **pinnate**, feathery.

poikilohydry—Having an interior water content that varies with external moisture changes; contrast with **homoiohydry**.

polysetous—With multiple **setae** per **perichaetium**.

propagule—A structure or fragment capable of reproducing a plant.

prorulate—Refers to cells with overlapping ends, causing one or both ends of the cell to be raised above the plane of the rest of the cell. Like cellular **papillae**, this characteristic can render a leaf surface dull.

protonema, pl. protonemata—The beginning growth from a bryophyte spore or other **propagule**, green and often filamentous.

pseudoparaphyllium, pl. pseudoparaphyllia—Resembling **paraphyllia**, small **filiform** to leaflike structures found at the bases of branches or branch buds.

pseudopodium, pl. pseudopodia—A false **seta**. In mosses this is usually a stalk with **gemmae** (see *Aulacomnium* spp.). In Sphagnaceae and Andreaeaceae, it is the stalk that elevates the **capsule**. It is a haploid **gametophyte** structure versus a true **seta**, which is diploid and part of the **sporophyte**.

pyriform—Pear-shaped.

ranked—Aligned neatly in rows. Usually applied to leaves on a branch or stem. See *Sphagnum rubellum*.

receptacle—Structure in liverworts supporting male and female organs. These are often elevated structures in Marchantiopsida, the complex liverworts.

recurved—Curved back or down. Used here with reference to leaf margins or tips.

reflexed—Curved back. Used here primarily with reference to **capsule** teeth.

rhizoids—Filamentous structures acting in moisture retention or anchoring a bryophyte to a substrate; see also **tomentum**.

rugose—Refers to leaves with transverse wrinkles or pleats.

serrate—Refers to leaves with fine teeth pointing toward the leaf tip.

seta, pl. setae—**Capsule** stalk in mosses and liverworts (see **pseudopodium** for exceptions). See illustration at **capsule**.

spinulose—With small spiny teeth.

splash cup—Cuplike structure made from modified leaves and holding **antheridia** (see Polytrichaceae for examples) or **gemmae** (*Tetraphis pellucida*).

sporophyte—The diploid generation of a bryophyte, it produces and distributes haploid spores. The sporophyte remains attached to, and is dependent on, the **gametophyte**.

squarrose—Refers to leaves bent away from the stem at approximately a 90-degree angle. See *Sphagnum squarrosum*.

stellate—Star-shaped.

stoloniferous—Having or producing **stolons**.

stolon—Creeping, often sparsely leaved stem, often with **rhizoids** anchoring the plant.

striate—Having lengthwise markings.

striolate—Having fine lengthwise markings.

strumose—Refers to a moss **capsule** with an asymmetrical enlargement (struma) at its base that resembles an Adam's apple or goiter. See *Oncophorus wahlenbergii*.

subpercurrent—Refers to a moss **costa** that stops short of the leaf tip; see **alar cells**.

subpinnately—Not as neat or regular as **pinnate**.

subtubulose—Refers to leaves rolled longitudinally into almost a tube shape.

subulate—Awl-shaped.

succubous—In a dorsal view, with leaves overlapping such that a leaf lies under the next leaf toward the tip of a branch; contrast with **incubous**.

sulcate—Deeply grooved or pleated lengthwise.

thalloid—A flat, relatively undifferentiated plant body. See hornworts and thalloid liverworts for good examples.

tomentose—Refers to leaves having **tomentum**.

tomentum—A woolly covering of **rhizoids**.

tracheophyte—Plants with a well-developed vascular system.

transverse—In liverworts, refers to leaves attached at a 90-degree angle to the stem length, neither **incubous** nor **succubous**.

truncate—With the tip appearing to be chopped off.

type specimen—The original specimen used to describe a species.

undulate—Wavy with horizontal wrinkles.

Annotated References

In case this book has stimulated your interest in further study of bryophytes the following list of references may help you expand your understanding of this fascinating group of plants.

References Requiring Only a Hand Lens or Dissecting Microscope

Glime, Janice M. 1993. The Elfin World of Mosses and Liverworts of Michigan's Upper Penninsula and Isle Royale. Houlton, MI: Isle Royale Natural History Association.
A very good beginner's book giving excellent information about bryophytes, bryophyte ecology, and bryophyte biology. It describes many species that occur throughout the Northeastern United States and adjacent Canada.

Ireland, Robert, and Gilda Bellolio-Trucco. 1987. Illustrated Guide to Some Hornworts, Liverworts and Mosses of Eastern Canada. Ottawa: National Museums of Canada.
Out of print, but you may be able get a pdf of the book for free by sending an email to Questions@mus-nature.ca. This surprisingly complete book provides identification keys and illustrations allowing identification of most of the bryophytes you are likely to encounter in the Northeast without requiring a compound microscope. Excellent keys, illustrations, and information on how to collect and study bryophytes.

Ley, Linda M., and Joan M. Crowe. 1999. An Enthusiast's Guide to the Liverworts and Hornworts of Ontario. Thunder Bay, Ontario: Lakehead University.
Excellent drawings and good coverage throughout our area. Available from the Claude Garton Herbarium, Lakehead University.

Lincoln, Mary S. G. 2008. Liverworts of New England: A Guide for the Amateur Naturalist. Memoirs of The New York Botanical Garden, vol. 99. Bronx: New York Botanical Garden Press.
Terrifically user friendly with photographs, drawings, and creative keys.

McKnight, Karl B., Joseph R. Rohrer, Kirsten McKnight Ward, and Warren Perdrizet. 2013. Common Mosses of the Northeast and Appalachians. Princeton, NJ: Princeton University Press.
Good coverage for the "true mosses," limited Sphagnaceae coverage, and does not include liverworts or hornworts. Keys are based largely on leaf shape and structure.

McQueen, Cyrus B. 1990. Field Guide to the Peat Mosses of Boreal North America. Hanover, NH: University Press of New England.
Sadly, out of print, but occasionally available from used book sellers. This is an interesting attempt to identify *Sphagnum* species using ecology and macroscopic characteristics in a random-access key that is a welcome break from the dichotomous keys usually encountered.

Munch, Susan. 2006. Outstanding Mosses and Liverworts of Pennsylvania & Nearby States. Sunbury Press.
Lovely photographs and descriptions of bryophytes allowing the identification of many of our more charismatic species. Can be ordered from the author at susanm@alb.edu, or from Amazon.com.

Technical References Requiring Compound Microscope Work

The idea of compound microscope work should not be daunting. Frequently, all that's required is a simple slide of a flat leaf.

Allen, Bruce. 2005. Maine Mosses, vol. 1. Bronx: New York Botanical Garden Press.
Along with volume 2, the most up-to-date of all references. Following taxonomic organization, this volume covers most of the acrocarps other than the Polytrichaceae. Authoritative and highly technical.

Allen, Bruce. 2014. Maine Mosses, vol. 2. Bronx: New York Botanical Garden Press.
Along with volume 1, the most up-to-date of all references. Also organized taxonomically, this volume primarily covers the pleurocarps, but it also includes the Polytrichaceae and a few acrocarp families. Authoritative and highly technical.

Andrus, Richard E. 1980. Sphagnaceae (Peat Moss Family) of New York State. Bulletin 442. Albany: New York State Museum.
Very helpful technical keys and illustrations of the challenging Sphagnaceae family.

Conard, Henry S., and Paul L. Redfearn. 1979. How to Know the Mosses and Liverworts. Dubuque, IA: W. C. Brown.
Available inexpensively from used book dealers, but with minimal illustration and out-of-date nomenclature.

Crum, Howard. 1991. Liverworts and Hornworts of Southern Michigan. Ann Arbor: University of Michigan Herbarium.
A very helpful introduction to liverworts and hornworts. Available from the herbarium.

Crum, Howard. 2004. Mosses of the Great Lakes Forest. Ann Arbor: University of Michigan Herbarium.
This is the most affordable technical reference to the moss flora of the Northeast. It misses a few of our Eastern mosses, particularly coastal species, but all in all, it's a great introduction to mosses and to Howard Crum's wit and wisdom.

Crum, Howard, and Lewis Anderson. 1981. Mosses of Eastern North America. New York: Columbia University Press.
This complete and impressive two-volume set is quite expensive, so it probably won't be your first purchase. But if you study bryology long enough, you'll buy it sooner or later. The nomenclature is out-of-date. Hard to believe, but for a scientific tome, it's quite readable. Missing are keys to families or genera.

Flora of North America, volume 27
Available online (and in print), and includes descriptions and range maps for most acrocarps. Volume 28 will complete the acrocarps and add the pleurocarps, and volume 29 will include the Liverworts. The descriptions and illustrations are first rate. Available at: http://fna.huh.harvard.edu/.

Grout, A. J. 1903. Mosses with Hand-Lens and Microscope: A Non-technical Handbook of the More Common Mosses of the Northeastern United States. Brooklyn, NY: Published by the author.
Grout published the first of his nontechnical moss guides in 1900 and continued revising these books until his death in 1947. Nomenclature is very out-of-date. The foremost bryologist of his day, Dr. Grout also published a three-volume technical treatise covering mosses of North America.

Grout, A. J. 1924. Mosses with a Hand-Lens: A Popular Guide to the Common or Conspicuous Mosses and Liverworts of the United States and Canada. Newfane, VT: Published by the author.

Ireland, Robert. 1982. Moss Flora of the Maritime Provinces. Ottawa: National Museums of Canada.
Still in print, and available from the Canadian Museum of Nature. This is a user-friendly technical manual, particularly complete for New England, and with excellent keys. Nomenclature is out-of-date.

Schuster, Rudolf M. 1966–1992. The Hepaticae and Anthocerotae of North America, in six volumes. Volumes 1–4 were published by Columbia University Press, and volumes 5 and 6 were published by the Field Museum of Natural History.
This hard to find set of six large volumes is an amazingly complete piece of work. Very technical.

Bryophyte Texts

Schofield, W. B. 2001. Introduction to Bryology. Caldwell, NJ: Blackburn Press.
Vanderpoorten, Alan, and Bernard Goffinet. 2009. Introduction to Bryophytes. Cambridge: Cambridge University Press.

Miscellaneous

Kimmerer, Robin Wall. 2003. Gathering Moss. Oregon State University Press. Corvallis, Oregon.
This is a compelling selection of essays. You can't read this book and not want to learn more about moss.

Malcom, Bill and Nancy. 2006. Mosses and other Bryophytes, an Illustrated Glossary. Micro-Optics Press, Nelson, New Zealand.
Can be purchased from the online store of the California Native Plant society. This is a superbly illustrated glossary and I recommend it highly.

Stuber, Stephanie. 2012. The Secret Lives of Mosses: A Comprehensive Guide for Gardens.
Published by the author, and available as a Kindle book from Amazon or as an ebook or paperback from Lulu (Lulu.com). A very helpful introduction to moss gardening with a focus on public gardens.

Tuba, Zoltan, Nancy G. Slack, and Lloyd R. Stark. 2011. Bryophyte Ecology and Climate Change. Cambridge: Cambridge University Press.
This impressive collection of articles from more than 50 authors is at times quite readable and at times very technical. It provides extensive and timely information about the importance of studying bryophytes.

Websites

These are really too numerous to get into, but I'll steer you toward a few of my favorites and leave you on your own to explore the surprisingly rich realm of bryophyte-related websites.

Consortium of North American Bryophyte Herbaria. http://bryophyteportal.org/portal/collections/index.php.
A site to access herbarium information from many large North American herbaria. See Introduction.

Glime, Janice M. Bryophyte Ecology. An e-book sponsored by Michigan Technological University and the International Association of Bryologists. http://www.bryoecol.mtu.edu/.
Bookmark this site. It's well written, packed with ecological information about bryophytes, and constantly being updated.

Michael Lüth's Image Gallery of Bryophytes. http://www.bildatlas-moose.de/.
An online collection of images of all bryophytes growing in Germany, it surprisingly covers almost all our northeastern North American species. A great resource.

Tropicos. http://www.tropicos.org.
This website, created by the Missouri Botanical Garden, provides up-to-date nomenclature and literature references for Bryophytes.

Index

Page numbers in **bold** indicate species accounts.